SPEAK·TRUTH·TO·POWER

CLAUSEWITZ'S TIMELESS TRINITY

Military Strategy and Operational Art

Edited by Professor Howard M. Hensel, Air War College, USA

The Ashgate Series on Military Strategy and Operational Art analyzes and assesses the synergistic interrelationship between joint and combined military operations, national military strategy, grand strategy, and national political objectives in peacetime, as well as during periods of armed conflict. In doing so, the series highlights how various patterns of civil–military relations, as well as styles of political and military leadership influence the outcome of armed conflicts. In addition, the series highlights both the advantages and challenges associated with the joint and combined use of military forces involved in humanitarian relief, nation building, and peacekeeping operations, as well as across the spectrum of conflict extending from limited conflicts fought for limited political objectives to total war fought for unlimited objectives. Finally, the series highlights the complexity and challenges associated with insurgency and counter-insurgency operations, as well as conventional operations and operations involving the possible use of weapons of mass destruction.

Also in this series:

British Generals in Blair's Wars
Edited by Jonathan Bailey, Richard Iron and Hew Strachan
ISBN 978 1 4094 3735 2

Britain and the War on Terror
Policy, Strategy and Operations
Warren Chin
ISBN 978 0 7546 7780 2

Confrontation, Strategy and War Termination
Britain's Conflict with Indonesia
Christopher Tuck
ISBN 978 1 4094 4630 9

Joining the Fray
Outside Military Intervention in Civil Wars
Zachary C. Shirkey
ISBN 978 1 4094 2892 3

Russian Civil-Military Relations
Robert Brannon
ISBN 978 0 7546 7591 4

Clausewitz's Timeless Trinity

A Framework For Modern War

COLIN M. FLEMING
University of Edinburgh, UK

ASHGATE

Published by
Ashgate Publishing Limited
Wey Court East
Union Road
Farnham
Surrey, GU9 7PT
England

Ashgate Publishing Company
110 Cherry Street
Suite 3-1
Burlington, VT 05401-3818
USA

www.ashgate.com

British Library Cataloguing in Publication Data
A catalogue record for this book is available from the British Library

The Library of Congress has cataloged the printed edition as follows:
Fleming, Colin M.
  Clausewitz's timeless trinity : a framework for modern war / by Colin M. Fleming.
    pages cm. -- (Military strategy and operational art)
  Includes bibliographical references and index.
  ISBN 978-1-4094-4287-5 (hardback) -- ISBN 978-1-4094-4288-2 (ebook)
  -- ISBN 978-1-4094-7398-5 (epub) 1. Clausewitz, Carl von, 1780-1831. Vom Kriege 2. Military art and science--Case studies. 3. Yugoslav War, 1991-1995. I. Title.
  U102.C6643F54 2013
  355.02--dc23
                                                                        2013014530
ISBN 9781409442875 (hbk)
ISBN 9781409442882 (ebk – PDF)
ISBN 9781409473985 (ebk – ePUB)

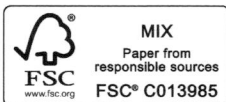

MIX
Paper from
responsible sources
FSC
www.fsc.org   FSC® C013985

Printed in the United Kingdom by Henry Ling Limited,
at the Dorset Press, Dorchester, DT1 1HD

To my lovely wife, Clairey

Contents

Acknowledgements

I would like to thank the Max Weber Programme at the European University Institute (EUI) for providing me with the intellectual stimulation and support while writing this book. There are numerous people who also I need to thank. I am especially grateful to Pascal Vennesson for providing me with the opportunity to present and gain feedback from his class at the EUI. I would also like to take the opportunity to thank my own students at Edinburgh University, to whom I presented my ideas. Their thoughts and comments have added to my own insights into Clausewitz and have made me reflect on my key arguments in this book. There are several other people I need to thank and who at different times have either proofread or provided ideas to my work. In particular, I would like to thank Michael Williams, Steven Haines, Mikkel Vedby Rasmussen and Christopher Coker; as well as David Barnes at the European University Institute, whose input was invaluable. Thank you also to Jim Wyllie, who has provided me with greatly appreciated support, as well as the time taken to read earlier drafts of this work. Lastly, I owe a special thank you to my wife, Claire, brother, Malcolm, dad, Jim, and mother, Margaret, and my good friend Lachie Nicolson, all of whom now know a great more about Clausewitz than they ever thought they would.

Introduction

This is the first book to apply the Clausewitzian Trinity of 'passion, chance, and reason' to the experience of real war. It explores the depth and validity of the concept against the conflicts of former Yugoslavia (1991–95) – wars thought to epitomize a post-Clausewitzian age. In doing so it demonstrates the timeless message of the Trinitarian idea. However, in a major departure from recent scholarship, it ties the Trinitarian idea back into Clausewitz's political argument.[1]

The ideas of Carl Von Clausewitz are enjoying a modern renaissance; the study of Clausewitz is flourishing. In many ways this is not surprising. Published posthumously in 1832, Clausewitz's *On War* has influenced strategic theory more than any other single work. Yet this influence has been balanced by controversy. Influential or not, his incomplete treatise has generated a great deal of spilt ink. Our present era is no exception. In the post-Cold War period, Clausewitz's ideas came under stinging criticism from scholars who associated his approach solely with inter-state war. In a strategic environment where this mode of war is the exception rather than the norm, his ideas have been frequently attacked. The 'new wars' of our contemporary era were un-Clausewitzian; war was no longer motivated by political rationality, but by ethnic hatred, religious fanaticism, and economic greed. Clausewitz, we were told, was obsolete; his ideas redundant in a more complex strategic age.

Best known for his dictum that 'war is the continuation of politics' Clausewitz's ideas have been closely tied to the idea that war could be a rational instrument of state policy. As Clausewitz argued, 'war is not merely an act of policy but a true political instrument, a continuation of political intercourse, carried on with other means'.[2] For Clausewitz, the symbiotic relationship between war and politics stems from the very essence of what conflict is – it plays a vital and functional role. As he is at pains to remind readers, war is 'an act of violence intended to compel our opponents to fulfil our will'.[3] From this starting position he understood war as a continuation of policy, driven by political objectives. A cornerstone of strategic studies throughout the Cold War, this idea appeared shattered as a wave of 'new wars' ripped up the rule book during the 1990s. Thought to be symbiotically linked

---

1   The book primarily utilizes the Howard and Paret translation. Although each translation has faults, this is the most widely used and accessible translation to date. See: Carl von Clausewitz (1882), *On War*, transl. by Michael Howard and Peter Paret (New York: Alfred A. Knopf, Everyman's Library edition, 1993).

2   Clausewitz, *On War*, 99.

3   Clausewitz, *On War*, 99.

to Clausewitz's supposed rational calculus, his 'Wondrous Trinity' came in for particular censure.

His Trinitarian concept is composed of the interactive elements of passion, chance, and policy, and was designed as an explication of war's complexity. Throughout the Cold War, however, its meaning had been misinterpreted. Clausewitz also discussed a second Trinity of people, army, and government. Thought to embody Clausewitz's understanding of rationality, this second Trinity was treated as confirmation of Clausewitz's state-centric approach. Getting the three elements of the triad – people, army, government – into alignment was interpreted as Clausewitz's secret formula for success. In a new strategic milieu, where inter-state war was going out of fashion, Clausewitz's Trinity was rubbished, supposedly outdated in a new era of strategic interaction.

These ideas have been successfully challenged, and the Trinity 'reclaimed'.[4] Clausewitz's modern supporters have successfully emphasized the core concept over the people, army, and government model. Writers such as Christopher Bassford, Antulio Echevarria, and Andreas Herberg-Rothe have all placed the Trinity at the heart of Clausewitz's theory of war; his 'legacy' as Herberg-Rothe puts it. Yet, in their defence of Clausewitz's Trinitarian model they have uncovered a major tension at the heart of his theory. Although Clausewitz is best known for his claim that war is instrumental, 'a continuation of politics', in his Trinitarian analysis war's political rationale is subsumed by the competing elements of the formula. The non-linear aspects of war that the Trinity represents appear to delimit Clausewitz's core political/instrumental message.

This book examines this anomaly. Although the Trinity is instructive and equips readers of *On War* with a deeper comprehension of war's complexity, the disjuncture from the political unit which gives it life is misleading and there is a danger that the Trinity in its current form becomes no more than yet another prescriptive model: the nature of war as a three-part formula. This is exactly what Clausewitz opposed and although war should be conceptualized as interactive, it must remain tied to context, and that is politics.

Intended to build on the current corpus of scholarship, it differs from the existing literature in two ways. By applying the Trinity to the wars of former Yugoslavia 1991–95, it explores war at its micro-foundations, assessing the complex cause-and-effect nexus of reciprocity produced by actions between belligerents embroiled in dynamic competition perpetuated by their own interaction. Providing valuable insights into the complexities of real war fuelled by passion, undermined

---

4   Important works include: Christopher Bassford, 'Tip-Toe Through the Trinity, or The Strange Persistence of Trinitarian warfare', www.Clausewitz.com/CWZHOME/ TRINITY/TRINITY8.htm (2007) (accessed 11 July 2009). An earlier version is published as: Christopher Bassford, 'The Primacy of Policy and the 'Trinity' in Clausewitz's Mature Thought', in Hew Strachan and Andreas Herberg-Rothe (eds), *Clausewitz in the Twenty-first Century* (Oxford: Oxford University Press 2009); Antulio J. Echevarria II, *Clausewitz and Contemporary War* (Oxford: Oxford University Press, 2007).

by chance, shaped by reason, it is the first study to bridge the Clausewitzian world of theory with real experience. Examining each part of the triad separately, the book explores the multiple manifestations of hostility and chance, before then tracing the influence of these elements on the policies of the belligerents as the war evolved.

As noted above, in a major point of departure from existing scholarship, the book also challenges the claim that the Trinity supersedes Clausewitz's political argument. The components of the triad are not necessarily 'a priori' equal, as recent works suggest.[5] Clausewitz tells readers that the 'three tendencies (of the Trinity) are like three different codes of law, deep-rooted in their subject and yet variable in their relationship to one another. A theory that ignores any one of them or seeks to fix an arbitrary relationship between them would conflict with reality.'[6] This position has brought recent scholarship to suggest the limitations of policy within Clausewitz's work. Yet, does this statement mean that the three elements are to be considered equal? Is policy subsumed by the Trinity? Although Clausewitz reminds readers not to fix an arbitrary relationship between the three components, he repeatedly explains that war is guided by policy. Through a critical interpretation of *On War* juxtaposed with the findings of the case study this claim is substantiated. Too much emphasis on the non-linear elements of Clausewitz's work, especially the interactive nature of the Trinity, undermines his core political argument. Clausewitz cautioned against favouring one element of the model over another. However, the reality of war, and the rest of his writings, suggests that policy remained the predominant element of the trinity.

This is important. Although the position is subtle, it is critical nonetheless, having a direct bearing on our ability to better understand contemporary war. Clausewitz reminds us of war's volatility by way of his model, but we must not lose sight of the fact that the Trinity is itself formed by the competition of opponents, and their political interests and aspirations hold the key to understanding war and the Trinity that it produces. Realigning the Trinity to better link with Clausewitz's wider ideas related to politics provides a better explanatory model of war today, grounding the utility of the Trinitarian concept to political context, and forcing analysts to understand the political, social, and cultural environment in which conflict takes place.

The book cuts across history, international relations, and strategic and security studies. Much of the literature and debate on strategic affairs since the Cold War has sought to explain changes in the practice of war, emphasizing difference and a disconnection from the rest of our strategic history. For many analysts of strategic affairs, the conclusion of the Cold War gave way to an era of 'new wars', which we still fail to understand. A problem exacerbated, it is argued, by the refusal to reject outdated theories – such as Clausewitz's *On War*. The book is set against

---

5   Antulio J. Echevarria II, *Clausewitz and Contemporary War* (Oxford: Oxford University Press, 2007), 73.

6   Clausewitz, *On War*, 99.

these debates, particularly the challenges posed to Clausewitz's ideas by the 'new war' thesis and the Revolution in Military Affairs (RMA).

These competing paradigms provide the essential context for the study. As illustrated below, recent challenges do not delimit the validity of Clausewitzian thought. In fact, far from being obsolete, Clausewitz's ideas are very much alive and continue to provide an explanatory foundation for our understanding of war in the twenty-first century. In fact, in an age where war is increasingly complex, Clausewitz's ideas are more rather than less powerful; and better able to capture the complexity that modern war inheres. Although much of the recent strategic and security studies literature focuses on how war is waged, on how war's character (the way it is fought) has been revolutionized, this book cuts deeper, exploring war at its interactive micro-foundations. Exploring the strengths and weaknesses of the Trinity, it balances the changes in war's evolving character with the continuity of its timeless nature, a reminder that war is 'still' war despite the many forms it can take. Neither the RMA nor the counter-RMA, which pits irregular insurgency and terrorism against the technological wizardry of the US and its allies, will change this essential fact. It is a fact that Clausewitz would be well aware of. By realigning the Trinity to account for the hierarchal position of policy/politics within the construct, we are left with a more powerful conceptual model which is better placed to make sense of the increasing complexity of modern war.

Clausewitz after the Cold War – The RMA and the 'New War' Challenge

The post-Cold War criticism of Clausewitz was widespread. The receding threat of nuclear war produced hopes of a 'peace dividend', a renewed sense of cosmopolitanism generated by economic globalization, and a willingness to deal with international crises through stronger ties to international organizations such as the United Nations (UN) and the European Union (EU). Coupled with the strengthening of international law, these factors appeared to be transforming international politics. For many, these changes also represented a major transformation in the nature of war, resulting in widespread calls to review our understanding of conflict for a new age. Some even suggested that in an increasingly globalized world, major war between major states was obsolete.[7] By the late 1990s commentators such as Michael Mandelbaum were claiming that the trend towards obsolescence had accelerated dramatically. As he argued, 'the rising costs of war, and the diminishing expectations of victory's benefits, have transformed its status'.[8]

It is now reasonable to question these assumptions, and as Colin Gray has noted, war has a healthy future. Arguments that war in the post Cold War era had

---

7   John Mueller, *Retreat from Doomsday: The Obsolescence of Major War* (New York: Basic Books, 1989).

8   Michael Mandelbaum, 'Is Major War Obsolete?' *Survival* (Winter, 1998–99), 1–2.

been 'transformed' have become particularly vocal. Although this transformation refers to war's character (how it is fought), it has been regularly conflated to include the transformation of war's nature (what war is); a mistake that continues to engender controversy and some confusion. The expected 'peace dividend' gave way to a series of conflicts across the globe during the 1990s; the post-Cold War era, we were told, was very different from our strategic past. War, it was argued, was undergoing radical transformation. For many, it was postulated that this transformation had even altered the nature of war itself. Consequently, it would be prudent to rethink our approach to war in the modern world. The terrorist attacks on New York and Washington on 11 September 2001, the resulting War on Terror, and the attendant wars in Iraq and Afghanistan have further reinforced this position.

Misinterpreted as a state-centric theorist, assaults on Clausewitz's modern validity are widespread. Two ideas, the so-called 'new war' thesis and the (American) Revolution in Military Affairs (RMA) compete to replace Clausewitz as the formative theory of contemporary war.

The New War Challenge

While the new war argument varies its primary claim is that modern conflict differs from its historical antecedents in three major ways: (i) structure; (ii) methods; and (iii) motives. Focusing chiefly on the rise of non-state war, the new war idea was predicated on the notion that the state was in terminal decline – a position that appeared to be supported by statistical evidence. According to one study of the period 1990–97, there were 24 intra-state conflicts. During the same period there was just one inter-state war: the 1991 Gulf War.[9] As the International Commission on Intervention and State Sovereignty (ICISS) reported in 2001, 'The most marked security phenomenon since the end of the Cold War has been the proliferation of armed conflict within states'.[10] The increasing use of irregular warfare by terrorist organizations and weaker powers is also claimed to loosen the bonds between state and military, thus accelerating and exacerbating the original problem of state decline.

This trend was highlighted as early as 1991; however, the perception that conflict in the new millennium ostensibly fulfils the 'new war' prophesy feeds into the idea that the nature of war is undergoing a transformation. Writing in 1991, military historian Martin Van Creveld argued that 'armies will be replaced

---

9   According to the Correlates of War data sets, there have been 79 inter-state wars between the years 1816–1997. During the same period there have been a total of 213 intra-state wars. See: Meredith Reid Sarkees, 'The Correlates of War Data on War: An Update to 1997', *Conflict Management & Peace Sciences* (2000), 1811: 123–44.

10   The Responsibility to Protect: Report of the International Commission on Intervention and State Sovereignty (Ottawa: International Development Research Centre, 2001), 4.

by police-like security forces on the one hand and bands of ruffians on the other'.[11] It will be extremely difficult not only to recognize the kaleidoscope of belligerents engaged in these conflicts, but the distinction between war and peace will become increasingly blurred; it would be an era of 'constant conflict'.[12] Some commentators even suggested that using the term 'war' at all gives it a credibility that belies its unorganized character.[13] As Kalevi Holsti has argued,

> War has become de-institutionalized in the sense of central control, rules, regulations, etiquette, and armaments. Armies are rag-tag groups frequently made up of teenagers paid in drugs, or not paid at all. In the absence of authority and discipline, but quite in keeping with the interests of the warlords, 'soldiers' discover opportunities for private enterprises of their own.[14]

Mary Kaldor, perhaps the best known of the new war advocates, explains the difference inherent in new wars:

> In contrast to the vertically organized hierarchical units that were typical of 'old wars', the units that fight these wars include a disparate range of different types of groups such as paramilitary units, local warlords, criminal gangs, police forces, mercenary groups and also regular armies including breakaway units of regular armies. In organizational terms, they are highly decentralized and they operate through a mixture of confrontation and cooperation even when on opposing sides.[15]

During the 1990s, war in former Yugoslavia, Caucasus, and in parts of Africa seemed to substantiate the 'new war' claims with much-needed evidence. Kaldor even claimed that 'the war in Bosnia-Herzegovina has become the archetypal example, the paradigm of the new type of warfare'.[16] Ostensibly, this argument is accurate. These conflicts do appear to exhibit irrational traits. These qualities also appear to shape governmental policy as much as simple political expediency. In Rwanda and in former Yugoslavia it is argued that ethnic hatred exacerbated the tense political context; it is also claimed that in Bosnia war was driven by criminal

---

11   Martin Van Creveld, *The Transformation of War* (New York; the Free Press, 1991), 225.

12   See: Ralph Peters, 'Constant Conflict', *Parameters* (Summer, 1997), 4–14; Mark Duffield, *Global Governance and the New Wars* (London, 2001), 161–201.

13   Monty Marshall, 'Systems at Risk: Violence, Diffusion, and Disintegration in the Middle East', in David Carmet and Patrick James (eds), *Wars in the Midst of Peace: The International Politics of Ethnic Conflict* (University of Pittsburg Press, 1997), 82–115.

14   Kalevi J. Holsti, 'The Coming Chaos? Armed Conflict in the Worlds Periphery', in T.V. Paul and John A. Hall (eds), *International Order and the Future of World Politics* (Cambridge: Cambridge University Press, 1999), 304.

15   Mary Kaldor, *New and Old Wars* (Cambridge: Polity Press, 2001), 8.

16   Kaldor, *New and Old Wars*, 31.

gangs intent on maintaining lucrative profit margins.[17] During many of the wars during the 1990s, traditional 'conventional' militaries rarely loomed large as the central players. As such, it has become common for commentators to envisage a world where 'conventional' armies cannot function properly against a new type of enemy; a trend that it was thought would accelerate as demographic problems were exacerbated by economic and environmental problems.

As we enter the beginning of the second decade of the twenty-first century, these arguments seem compelling. Despite US success in the 2007 'surge' in Iraq, US and coalition forces have struggled to adapt to an irregular foe which uses the urban environment as a force multiplier.[18] In Afghanistan, the Taliban have used the mountain environment to circumvent allied power and sustain provisions through the porous border crossings into Pakistan – a state which itself seems to teeter on the brink of civil war as its conventional forces struggle to deal with their one-time proxy – the Taliban. The 'summer war' in 2006 between Israel and Hezbollah, and the Israeli-Palestinian conflict during the winter of 2008/2009 further entrench this view.[19] As an 'asymmetric' opponent, the Taliban in Afghanistan has regrouped since its initial defeat in 2001. The US, UK, and the North Atlantic Treaty Organization (NATO) appear bogged down in a conflict that is increasingly hard to win; in the traditional sense at least.[20] As Christopher Coker reflects, to avert defeat in Afghanistan and Iraq, 'conventional' militaries have increased their own transformation towards smaller and more mobile force structures able to better locate and engage opponents.[21] This has been in addition to an increasing emphasis on counterinsurgency techniques and a reduction of kinetic force.

For writers such as William Lind (1989), Martin Van Creveld (1991), Mary Kaldor (2001), and Herfried Münkler (2005) amongst others, this unstructured non-state mode of conflict was evidence of the modern redundancy of Clausewitzian theory, for so long the cornerstone of traditional strategic studies.[22]

---

17   Important works on this topic include: Kaldor, *New and Old Wars*; Paul Collier, 'Doing Well Out of War: An Economic Perspective', in Mats Berdal and David M. Malone, *Greed and Grievance: Economic Agendas in Civil Wars* (Boulder: Lynne Rienner Publisher, 2000), 91–111; David Keen, 'The Economic Functions of Violence in Civil Wars', *Adelphi Paper*, 320 (London: International Institute of Strategic Studies, 1998).

18   David Kilcullen, *The Accidental Guerrilla* (London: Hurst & Co., 2009), 115–86. This problem is not restricted to the US or the coalition in Iraq.

19   Avi Kober, 'The Israel Defence Forces in the Second Lebanon War: Why the Poor Performance?', *Journal of Strategic Studies*, 31(1) (2008), 3–40.

20   See: M.J. Williams, *The Good War: NATO and the Liberal Conscience in Afghanistan* (Palgrave, 2011).

21   Christopher Coker, *War in an Age of Risk* (Cambridge: Polity Press, 2010), 161.

22   See: William S. Lind, 'The Changing Face of War: Into the Fourth Generation', *Military Review* (1989), 69; Martin Van Creveld, *The Transformation of War* (New York: The Free Press, 1991); and Kaldor, *New and Old Wars*. See also: Herfried Münkler, *The New Wars* (Cambridge: Polity Press, 2005).

Because the Trinity was supposedly linked to a rationalist means-end calculus of war – thought to comprise the tripartite of people, army and government, the foundations of the state and its ability to wage war – it came in for particular censure. Van Creveld argued:

> The threefold division into government, army, and people does not exist in the same form. Nor would it be correct to say that, in such societies, war is made by governments using armies for making war at the expense of, or on behalf of, their people.[23]

Claiming that the model represents the embodiment of inter-state war, he propounded the argument that war as a continuation of politics – as articulated by Clausewitz – is a misnomer in the modern world. War, Van Creveld, argued, would be 'driven by a mixture of religious fanaticism, culture, ethnicity, or technology'.[24] Under these conditions war could no longer be regarded as a 'political instrument', as Clausewitz famously suggested.

This particular Clausewitzian argument has been the cornerstone of traditional strategic studies theory. Clausewitz argues that despite war's violent proclivities, it must be governed by political objectives. War should be fought in the pursuit of political goals. War, he reminds readers, 'is not merely an act of policy but a true political instrument'. Reiterating the point, he extols that: 'The political object is the goal, war is the means of reaching it, and means can never be considered in isolation from their purpose'.[25]

Although misinterpreted, the idea that Clausewitzian theory embodies a rationalist means-end calculus was thought to be encapsulated in his 'Wondrous Trinity'. As noted above, this idea misinterprets Clausewitz's real argument. The locus of the attack on Clausewitz was founded on the mischaracterization of the Trinity as comprising people, army, and government. For new war theorists the concept is indicative of why Clausewitzian ideas are moribund. Clausewitz wrote that:

> War is more than a true chameleon that slightly adapts its characteristics to a given case. As a total phenomenon its dominant tendencies always make war a paradoxical trinity – composed of primordial violence, hatred, and enmity which are to be regarded as a blind natural force; of the play of chance and probability within which the creative spirit is free to roam; and of its element of subordination, as an instrument of policy, which makes it subject to reason alone.

He continues:

---

23 Van Creveld, *Transformation of War*, 49–50.
24 Van Creveld, *Transformation of War*, 69.
25 Clausewitz, *On War*, 99.

The first of these three aspects mainly concerns the people; the second the commander and his army; the third the government. The passions that are to be kindled in a war must already be inherent in the people; the scope which play of courage and talent will enjoy in the realm of probability and chance depends on the particular character of the army; but the political aims are the business of government alone.[26]

By marrying the Trinity to sections of society, many scholars have assumed that the concept is fundamentally linked to the state. As shown in more detail in the following chapter, the confusion surrounding the Trinity regards several interlinked factors. However, because this second Trinity, of 'people, army, and government', had taken precedence over Clausewitz's original construct – emotion, chance, and reason, it seemed to be misplaced in an age of non-state war. It is this 'second' Trinity that the new war writers censure. Because of the proliferation of non-state groups, the traditional state-centric political base for war is redundant – leaving the 'people' as the only remaining component of the Trinity. Accordingly, war would be stripped of its rational elements. As it is thought that the 'people' will form armed militias or mobs, which do not have structures able to promote rationality in the advance of their conflicting cultural, religious or ethnic aims, it is assumed that war can no longer be described as a rational political activity. Put simply, the rational elements of the Clausewitzian Trinity, the military and principally the government, are supposedly no longer present. Consequently, the appropriate 'rational' component of the concept cannot restrict the irrational traits that all wars exhibit. With the end of the state, and therefore the international system of states, only violent and 'non-Trinitarian', non-political war will remain.[27]

As noted above these ideas have been successfully challenged. Indeed, when one scrutinizes the new war thesis closely there is very little that can be labelled particularly new.[28] War, as Jeremy Black has pointed out: 'changes far less frequently and significantly than most people appreciate'.[29] Moreover, as I have noted elsewhere, the very fact that war evolves, that its character changes from one age to the next does not delimit Clausewitz's ideas. Changes in the character of war in the post-Cold world are not exclusive of Clausewitzian theory, which remains wonderfully placed to act as the touchstone of our strategic knowledge.

---

26   Clausewitz, *On War*, 101.

27   Non-Trinitarian war is a term coined by Van Creveld to express the redundancy of Clausewitz's original Trinity.

28   See: Colin M. Fleming, 'Old or New Wars? Debating a Clausewitzian Future', *The Journal of Strategic Studies*, 32(2) (April, 2009), 213–41.

29   Jeremy Black, *War in the New Century* (London: Continuum International Publishing Group, 2001), 114.

The fact that non-state war is part of the contemporary strategic and security milieu does not delegitimize Clausewitz's ideas, which sought to understand war's eternal nature rather than its temporal character. Clausewitz and his contemporaries were well aware that war existed beyond the bounds of the state. Clausewitz dedicated an entire chapter in Book Six of *On War* to this very issue, and lectured on the subject of 'small wars' at the *Berliner Kriegsschule* from 1811–12.[30] Jomini, who had participated in a guerilla war himself, even wrote that:

> As a soldier, preferring loyal and chivalrous warfare to organized assassination, if it be necessary to make a choice, I acknowledge that my prejudices are in favour of the good old times when the French and English Guards courteously invited each other to fire first, – as at Fontenoy, – preferring them to the frightful epoch when priests, women, and children throughout Spain plotted the murder of isolated soldiers.[31]

The classical war thinkers may have desired a return to an era when conflict did not deviate from the strict parameters imposed by the ruling elites of the eighteenth century, yet they were acutely aware of other modes of warfare. As Clausewitz made clear in Chapter Three, Book Eight, every age has its own kind of war, 'its own limiting positions, and its own peculiar preconceptions'.[32] He reflects,

> The semi-barbarous Tartars, the republics of antiquity, the feudal lords and trading cities of the Middle ages, eighteenth-century kings and the rulers and peoples of the nineteenth century – all conducted war in their own particular way, using different methods and pursuing different aims.[33]

Clausewitz's assessment of war throughout history illustrates his awareness that the character of war was constantly changing, often dramatically, from one age to the next. The very point that he was making was that despite war's evolving character, its special nature is universal. Returning once more to his Trinity, he reminds readers that: 'War is more than a true chameleon that slightly adapts its characteristics to a given case. As a total phenomenon its dominant tendencies always make war a paradoxical trinity'.[34] By emphasizing that war is 'more that a chameleon', Clausewitz informs us that war's nature should not be confused with the way it looks. That it alters its appearance and character 'to a given case' is unimportant. The unifying element that ensures the universality of the nature of

---

30   Christopher Daase, 'Clausewitz and Small Wars', in Hew Strachan and Andreas Herberg-Rothe (eds), *Clausewitz in the Twenty-first Century*, 183.

31   Baron Antoine-Henri de Jomini (1862), *The Art of War*, edited with an introduction by Charles Messenger (London: Greenhill Books, 1992), 34–5.

32   Clausewitz, *On War*, 715.

33   Clausewitz, *On War*, 708–9.

34   Clausewitz, *On War*, 101.

war has nothing to do with the conduct of war. It relates directly to the fact that war's nature comprises the interplay of the different elements within his Trinitarian concept: hostility, chance, and reason/policy.

In any case, thinking that irregular conflict is somehow a modern phenomenon, which must recast our understanding of the nature of war, is a mistake. One thinks of Thomas Hammes' argument that the architects of Fourth Generation Warfare (4GW) 'convince the enemy's political decision makers that their strategic goals are either unachievable or too costly'.[35] As a strategy, this hardly seems revolutionary. In his *On Guerrilla War* Mao Tse-tung argues that protracted conflict is a critical stage in his 'three-phased war'.[36] Furthermore, as Beatrice Heuser explains, 'Clausewitz made quite detailed prescriptions for the use of the *guerrilla*'.[37] She comments:

> He realised that while the best way to victory is unquestionably to have larger armies and to defeat a smaller or weaker enemy army utterly in one main battle, other factors can favour smaller or weaker powers. Apart from morale, this could be a greater stamina and patience, so that a larger enemy might not be prepared to invest the same amount of time to a particular conflict as the weaker force.[38]

By employing asymmetric, irregular, warfare as a mode of fighting a technologically or quantitatively superior opponent, a belligerent is subject to the same strategic logic as its 'conventional' opponents. As noted already, although the characteristics of such a conflict will be different to war between states, it is unclear why this should delimit war's political nature. Moreover, it is clear that Clausewitz's principles relating to 'small wars' are evident in the tactics employed by forces in those contemporary conflicts now described as 'new':

> A general uprising, as we see it, should be nebulous and elusive; its resistance should never materialize as a concrete body, otherwise the enemy can direct sufficient force at its core, crush it, and take many prisoners ... On the other hand, there must be some concentration at certain points: the fog must thicken and form a dark and menacing cloud out of which a bolt of lightning may strike at any time.[39]

As to the claim that we have entered an era of 'constant conflict', where individual wars peter on without end, and where the distinction between peace and war is blurred, qualification is badly needed. War, as Clausewitz was aware,

---

35   Thomas X. Hammes, *The Sling and the Stone: On War in the 21st Century* (St Paul: Zenith Press, 2004), 2.

36   Mao Tse-tung, *On Guerrilla Warfare*, translated by Brig.-Gen. Samuel B. Griffith (1961) (Urbana: University of Illinois Press, 2000), 51–7.

37   Beatrice Heuser, *Reading Clausewitz* (London: Pimlico, 2002), 136.

38   Beatrice Heuser, *Reading Clausewitz*, 137.

39   Clausewitz, *On War*, 581.

is an extremely volatile activity – he was at pains to remind readers that it should not be entered into lightly. When war was joined, he cautioned, it should be the means of reaching a better political settlement. His warning to readers resonates loudly today:

> The ultimate outcome of a war is not always to be regarded as final. The defeated side often considers the outcome merely as a transitory evil, for which a remedy may still be found in political conditions at some later date.[40]

That war in the twenty-first century should also be regularly indecisive should not come as a great shock. Victory according to 'traditional' metrics of success with an emphasis on battlefield outcomes may be rarer than in the past; however, one should be reminded of the fallacy of thinking that victory in past wars was any more decisive than today.[41] As Brian Bond notes, 'all students of history must be struck by the ambivalence, irony, or transience of most military victories, however spectacular and decisive they appear at the time'.[42] Even during what is thought of as the epitome of Clausewitzian conflict in the nineteenth century, war frequently proved indecisive. It was a situation that would regularly be repeated during the twentieth century.[43] The notion that war has become unusually inconclusive rests with the idea that war is solely the domain of states. However, recent scholarship on perceptions of victory and defeat demonstrates that success remains possible, highlighting new measurements for success and failure rather than delimitating the idea of victory itself.[44]

Although it is perfectly conceivable that the state may lose its central status at some future juncture, Clausewitz's concept is as equally relevant to non-state warfare. Of course, this does not imply that the trends flagged by the new war writers are not worthy of attention. They highlight important developments which appear to be changing the conduct of modern conflict and these trends must be taken seriously. The very fact that non-traditional security concerns exacerbate traditional security calculations is in itself enough to warrant significant attention.

---

40   Clausewitz, *On War*, 89.

41   For important works on 'battlefield' victory, see: Stephen Biddle, *Military Power: Explaining Victory and Defeat in Modern Battle* (Princeton, N.J.: Princeton University Press, 2004); and, William Martel, *Victory in War: Foundations of Modern Military Policy* (Cambridge: Cambridge University Press, 2007).

42   Brian Bond, *The Pursuit of Victory: From Napoleon to Saddam Hussein* (Oxford: Oxford University Press, 1996), 1.

43   Michael Howard, 'When Are Wars Decisive?', *Survival*, 41(1) (Spring 1999), 126–35.

44   Dominic D.P. Johnson and Dominic Tierney, *Failing to Win: Perceptions of Victory and Defeat in International Politics* (Cambridge, MA: Harvard University Press, 2006); Robert Mandel, 'Defining Postwar Victory', in Jan Angstrom and Isabelle Duyvesteyn (eds), *Understanding Victory and Defeat in Contemporary War* (Abingdon: Routledge, 2007).

Focusing attention on non-state conflict, the authors of the new war idea have opened up a more complex strategic environment for analysis. At the very least, they have re-ignited debate about warfare in the modern world, so long shaped by the contours of the international state system. If one is truly going to grasp the complexity of war, then reflection of what factors are critical to understanding it is a positive step. Nonetheless, despite apparent prescience, many of the new war claims become exaggerated. Not least, the founding premise that it offers an insight into a non-Clausewitzian universe is unfounded. Clausewitz's core Trinity comprises the three tendencies: hostility, chance, and reason. It is not affected by state decline. It is for this same reason that Clausewitz's message of complexity is not forgotten. This is particularly true in relation to the still ongoing debate on military transformation spurned by the American RMA.

The RMA

For proponents of the technical revolution that spurned the ongoing American-led Revolution in Military Affairs (RMA), innovative technological advances seem to offer the promise of a new strategic reality, one in which Clausewitzian notions of 'chance' and 'friction' would be consigned to the historical rubbish heap.[45] Taking the technological RMA to its ultimate conclusion it is thought that war can truly become an instrument of state policy; stripping away the uncertainty and complexity that Clausewitz's *On War* communicates; making war truly a rational instrument of state policy.[46] Intimately associated with the related concept of 'military revolution', first mooted by historian Michael Roberts (1967), the RMA idea first appeared in the late 1970s when Soviet analysis predicted that US technological superiority would allow future 'conventional' firepower parity with tactical nuclear weapon capability.[47] The concept appeared to come of age during the 1991 Gulf War. In the aftermath of stunning US and allied military success, strengthening the qualitative advantage of the US and its allies was seen as central to maintaining and enhancing American military dominance.

---

45   For important works specifically linked to Clausewitz, see: Williamson Murray, 'Clausewitz Out, Computers In: Military Culture and Technological Hubris', *The National Interest* (Summer, 1997). See also: Steven Metz, 'A Wake for Clausewitz: Toward a Philosophy of 21st Century Warfare' *Parameters* (Winter, 1994–95), 126–32. For an excellent discussion of the utility of Clausewitzian thought in the information age see: David Lonsdale, 'Clausewitz and the Information Age', in Hew Strachan and Andreas Herberg-Rothe (eds), *Clausewitz in the Twenty-first Century* (Oxford: Oxford University Press, 2007), 231–50.

46   Mikkel Vedby Rasmussen, *The Risk Society at War: Terror, Technology and Strategy in the Twenty-first Century* (Cambridge: Cambridge University Press, 2006), 43–91.

47   Michael Roberts, 'The Military Revolution', *Essays in Swedish History* (London: Weidenfeld & Nicolson, 1967); see: Eliot A. Cohen, 'Change and Transformation in Military Affairs', *The Journal of Strategic Studies*, 27(3) (2003), 395–407.

Despite the proliferation of non-state opponents the US now fights, the mantra of transformation continues, premised on the idea that technological superiority is the key to US military success in the new millennium.[48] The obvious conclusion drawn from *Operation Desert Storm* was that by using 'smart' weapons the US could now hit an increasing number of targets worldwide. The application of technology – especially information technology – means that obstacles such as distance no longer restrict military decisions. Aided by superior information-based systems, C4ISTAR (Command, Control, Communications, Computers, Intelligence, Surveillance, Targeting, and Reconnaissance), the US military aims to overwhelm its enemies' ability to resist. Network Centric Warfare (NCW) dramatically speeds up the operational tempo by quickening the OODA loop (Observe, Orient, Decide, Act), a result of which is more effective targeting of the enemy system.[49]

In addition, one of the major advantages of the RMA is the ability of US forces to use technological information-based systems to radically reduce casualties, both to its own forces and to the enemy. This aspect of the RMA seems to suggest a quick fix to those who believe that the US suffers from a political sensitivity to casualties – the so called CNN effect. If, as Edward Luttwak suggests, the West has entered a 'post-heroic' age, where toleration of high numbers of casualties is no longer acceptable to populations not used to the deprivations of war, then the RMA offers a panacea for the US military.[50] Christopher Coker has even claimed that technology is the antidote to those who view war as either abhorrent, or indeed too risky. According to Coker's *The Future of War*, the industrialization of military conflict in the first half of the twentieth century has led to disenchantment with warfare. Re-enchantment will come with revolutions in information and biotechnology, which 'have invested it with a renewed lease of life'.[51]

Nevertheless, the problem with the most recent American RMA is that RMAs are most conspicuous during wartime, when one side reacts to a specific situation in order to gain an operational advantage. Essentially an RMA is the creation of a new way of destroying and defeating the enemy. The fact that the initial period of 'American RMA' based on the technological superiority of the United States

---

48   See: Donald H. Rumsfeld, 'Transforming the Military', *Foreign Affairs*, 81(3) (May/June, 2002), 20–32.

49   The British version of NCW is Network Enabled Capability (NEC), although the two are closely related. For information on Britain's own ongoing transformation, see: Theo Farrell, 'The Dynamics of British Military Transformation', *International Affairs*, 84(4) (July 2008).

50   Edward Luttwak, 'Toward Post-Heroic Warfare', *Foreign Affairs*, 74(3) (1995), 109–22. It has been postulated that changes in Western demographics and economic security have resulted in a de-bellicized society. Precision Guided Weapons can locate enemy positions without causing intolerable collateral damage or endangering one's own personnel; an obvious plus when a war effort can suffer from bad public relations. In contrast, in areas like the West Bank, Gaza, or indeed in Iraq or Afghanistan, death in battle is celebrated through martyrdom, and battle retains its Homeric appeal.

51   Christopher Coker, *The Future of War* (Oxford: Blackwell, 2004), xii.

has largely been achieved in a period of peace indicates an exception to the rule. Moreover, conceptually the RMA can suffer from similar problems to that of 'military revolutions'. With projects often taking decades to come to fruition, it is hard to argue that it is always a revolutionary phenomenon. When ostensibly ground-breaking transformation takes place over a number of years of relative peace, rather than subject to the military pressures one might expect, there is a case for re-examining whether an RMA has really taken place at all. Nevertheless, the RMA continues to attract important supporters and advocates of US military transformation, who remain convinced of the efficacy of technology as the answer to strategic success. Moreover, although the first round of the US RMA took place during a time of relative peace, the second phase – the 'robotics revolution' has been in response to the counter-RMA of insurgents pitting low-tech (improvised explosive devices, and suicide bombs) against high-tech US dominance in Iraq and Afghanistan.[52]

Throughout the debate, it has been suggested that the RMA can limit the impact of uncertainties such as 'chance' and 'friction', making war more straightforward, producing a cleaner battle-space, and making the prospect of military casualties more remote. If the RMA is to be believed then these elements of war have been overcome. The rejoinder to this claim is equally powerful: intricate computer systems and other equipment can in many cases actually exacerbate and magnify these very ambiguities – computers break down and opponents adapt. While American military power is unrivalled, its enemies rarely play to its strengths, and instead utilize unconventional, asymmetric strategies. As Lawrence Freedman has argued, the result is not simply a 'Revolution in Military Affairs', but a true 'Transformation of Strategic Affairs'.[53] In fact, while the wars in Iraq and Afghanistan have not resulted in US casualties on the same scale as earlier wars, they have hardly been 'clean' risk free affairs – in Iraq, US casualties are estimated at 4,270 between 2003 and April 2010; in Afghanistan the figure stands at an estimated 1,082 deaths confirmed between 2001 and 2010.[54]

---

52   See: P.W. Singer, *Wired for War: The Robotics Revolution and Conflict in the 21st Century* (London: Penguin, 2009). Other works include; David Axe and Steve Olexa, *War Bots: How US Military Robots Are Transforming War in Iraq, Afghanistan, and the Future* (Ann Arbor, Mich.: Nimble Books, 2008); Antoine Bousquet, *The Scientific Way of Warfare: Order and Chaos on the Battlefields of Modernity* (London: Hurst & Co., 2009).

53   Lawrence Freedman, 'The Transformation of Strategic Affairs', *Adelphi Paper*, 45(379) (March 2006). This trend is not solely associated with non-state actors. States are also examining how asymmetric force can delimit US material military power. Chinese strategists have been studying asymmetry as a means of combating US power since the 1990s. See: Quiao Lang and Wang Xiangsu, *Unrestricted Warfare* (Bejing: PLA Literature and Arts Publishing House, 1999). See also: Frank G. Hoffman, '"Hybrid Threats": Reconceptualising the Evolving Character of Modern War', *Strategic Forum*, Institute for National Strategic Studies. National Defense University. No. 240, (April, 2009), 1–8.

54   Iraq Index, The Brookings Institute; April 27, 2010, www.brookings.edu/ iraqindex, 14; Afghanistan Index, The Brookings Institute, May 28, 2010, www.brookings. edu/afghanistanindex, 11 (accessed 30 July 2010).

As the pioneering work on the linkages between evolutionary biology and war illustrates, combatants are quick to adapt to new hurdles, formulating new tactics and strategies which severely impede material strength.[55] In this scenario competing powers become locked into an adaptation arms race where evolutionary 'selection effects favour the weaker side, such as insurgents and terrorists, because they are more varied, and are under stronger selection pressure, and replicate successful strategies faster than larger forces trying to defeat them'.[56]

It is axiomatic that the practice of war by the strongest states will be built on the current technological/digital revolution. Not least, robotic systems such as Unmanned Aerial Vehicles (UAVs) and computerized remote control are appealing because they offer the promise of further reducing the severity of war and the risk of casualties. Nevertheless, as this book, and the Trinitarian formula, reminds us, war is an interactive activity not easily corralled by technology. That it is part of the fabric of war is not in doubt: it has driven earlier RMAs and resulted in real Military Revolutions. Yet, technology does not strip war of its essential elements. In fact, rather than simplifying war, focus on the 'technological means' of war serves to de-contextualize conflict from the social, cultural, and political context that it takes place in. War remains at its micro-foundations a contest between opposing wills. This interaction and complexity is illuminated by the Trinitarian idea – war naturally produces unpredictability. As Williamson Murray and MacGregor Knox warn, the RMA must be understood as something shaped by politics and doctrine, as well as technology.[57] Although the RMA concept is a useful one, it forms part of a larger political and strategic context. Truism or not, as Colin Gray points out, 'the person behind the gun matters more than the gun itself'.[58]

These arguments continue to resonate strongly today. As a result Clausewitz's ideas, framed by a political context and fed by war's natural reciprocity, must continue to act as the foundation of our strategic knowledge. The RMA may be real, but technology alone does not make a war. More particularly, a better appreciation

---

55   Rafe Sagarin, 'Adapt or Die: What Charles Darwin Can Teach Tom Ridge about Homeland Security', *Foreign Policy* (September/October, 2003), 68–9; Dominic D.P. Johnson (2009), 'Darwinian Selection in Asymmetric Warfare: The Natural Advantage of Insurgents and Terrorists', *Journal of the Washington Academy of Sciences*, 95(3) (2009), 89–112; King, A.J., Johnson, D.D.P. and Van Vugt, M. (2009), 'The Origins and Evolution of Leadership', *Current Biology* 19(19), 1591–682. See also: Bradley A. Thayer, 'Bringing in Darwin: Evolutionary Theory, Realism and International Politics', *International Security*, 25(2) (Fall 2000), 124–51; and, Rafe Sagarin and Terry Taylor (eds), *Natural Security: A Darwinian Approach to a Dangerous World* (Berkeley, University of California Press, 2008).

56   Dominic D.P. Johnson (2009), 'Darwinian selection in asymmetric warfare: the natural advantage of insurgents and terrorists', *Journal of the Washington Academy of Sciences*, 95(3) (2009), 92.

57   MacGregor Knox and Williamson Murray, 'Thinking about Revolutions in Warfare', in MacGregor Knox and Williamson Murray (eds), *The Dynamics of Military Revolution, 1300–2050* (Cambridge: Cambridge University Press, 2001), 180–81.

58   Colin Gray, *Another Bloody Century* (London: Weidenfeld & Nicolson, 2005), 122.

of the ways in which the Trinity works, and of how it relates to Clausewitz's understanding of war as a political phenomenon, acts as counterargument to those hailing the totality of the technological digital revolution. War has a context in which it takes place. This is the message of *On War*, underpinned by the Trinitarian idea. As highlighted already, this means that each war must be treated as essentially different, with different prevailing conditions as it begins. The RMA can mask this reality and the advance of technology acts to artificially de-contextualize conflict from the culture and the social and political milieu in which it takes place; but it is only part of a wider political and strategic context that unfolds as the interaction itself unfolds.

The Scope of the Book

The following chapters are not intended as a continuation of the 'new' war/'old' war debate. Neither are they directly focused on the RMA proponents. The ideas promulgated against Clausewitz by the new war theorists have been successfully challenged. Clausewitz's modern supporters have successfully emphasized the core Trinity over the people, army, and government model. Writers such as Christopher Bassford, Antulio Echevarria and Andreas Herberg-Rothe have all placed the Trinity at the heart of Clausewitz's theory of war – his 'legacy' as Herberg-Rothe put it. They have revivified scholarship on Clausewitz as a consequence.

Yet, in their defence of Clausewitz's Trinitarian model they have uncovered a major tension at the heart of his theory. Although Clausewitz is best known for his claim that war is instrumental, 'a continuation of politics', in his Trinitarian analysis war's political rationale is subsumed by the competing elements of the formula. The non-linear aspects of war that the Trinity represents appear to delimit Clausewitz's core political/instrumental argument. This book examines this anomaly. As noted earlier, while intended to build on the existing corpus of scholarship it differs from the existing literature in two ways. Firstly, by applying the Trinity to the wars of former Yugoslavia (1991–95) it explores war at its micro-foundations, assessing the complex cause-and-effect nexus of reciprocity produced by belligerent interaction. Examining each part of the triad separately, the book explores the multiple manifestations of hostility and chance, before then assessing the influence of these elements on the policies of the belligerents as the war evolved. In doing so, it bridges traditional discourse with critical interpretation, offering new insights into Clausewitzian theory and the nature of war.

Secondly, as noted at the beginning of this introduction, in a major point of departure from existing scholarship, the book challenges the claim that the Trinity supersedes Clausewitz's political argument. The Trinity is truly a theory of modern war, but the non-linear elements of Clausewitz's theory should not overshadow the central importance of Clausewitz's political argument. Through a critical interpretation of *On War* tested against the experience of real war, this view is challenged. Although Clausewitz reminds readers not to fix an arbitrary relationship between the three tendencies of the Trinity, he repeatedly explains that war is guided by policy. A close reading of *On War* juxtaposed with the findings of the case study substantiates this claim. Too much emphasis on the non-linear elements of Clausewitz's work undermines his core political argument. Clausewitz cautioned against favouring one element of the model over another. However, the reality of war, and the rest of his writings, suggests that policy remained dominant in his thinking. The position is subtle, but it is critical nonetheless, having a direct bearing on our ability to better understand contemporary war. Clausewitz reminds us of war's volatility by way of his Trinitarian model, but we must not lose sight of the fact that the Trinity is itself formed by the competition of opponents, and their political interests and aspirations hold the key to understanding war and the Trinity that it produces.

Clausewitz's Trinity offers insights into contemporary problems, but to be truly useful there must be a willingness to refine his ideas to meet our own strategic setting. At a time when so much attention is focused on the changing character of war this is an important message. The Trinity alerts us to the complexity, interactivity, and unpredictability of war in any age and is of lasting value. In the post-Cold War era, strategic studies literature has understandably focused on changes in the character of war. These changes have indeed been profound. Yet, the RMA, and the increase in asymmetric warfare by non-state groups that is partly a result, do not mean the end of Clausewitzian theory. Clausewitzian theory illuminates the areas that contemporary theorists fail to reach. Of course, better understanding of changes in the character of war is crucial. Yet, just as change is a major factor in war, to truly understand contemporary conflict we must balance change with continuity. This book addresses the imbalance. It explores the foundations of the Trinitarian model by examining the ways in which hostility, chance, and rationality influence war, how they interact, and the consequences that follow that make war so difficult and dangerous an activity. These features are timeless – the micro-foundations that define war's nature rather than its evolving character. Clausewitz's Trinitarian model explicates war's dynamism, its dangers and opportunities.

Method and Structure: A Balkan Case Study?

The decision to examine Clausewitz's Trinity against the experience of the Balkan Wars in 1991–95 was made for several reasons. Unfortunately there is no shortage of wars to choose from. War in Iraq, Afghanistan, and the wider 'War on Terror' are probably the most widely known present examples, though there are many more contemporary conflicts which have attracted much less media time. These conflicts also offer interesting lines of enquiry and if Clausewitzian theory is to retain its modern relevance it must be able to inform and ameliorate our comprehension of war across its modern setting. It is intended that this current work will act as a stepping stone to a greater comprehension of these wars. Nevertheless, there are several interlinked reasons that make the wars of Yugoslavia particularly interesting.

Although the present conflicts in Iraq and Afghanistan are current and would thus provide an interesting test of the adaptability of Clausewitzian ideas to modern examples, the very fact that they are still ongoing significantly restricts the validity of the core question of this book, which seeks to determine the relationship between policy and the other tendencies in the Trinitarian concept. When war is ongoing, and therefore buffeted about by the different forces in the Trinity, gauging whether policy holds a predominant position is particularly difficult. Exploring the Trinity against a war from our recent past is much easier, providing time to pause and reflect which is difficult when wars are ongoing. It is possible to chart the journey of policy from the beginning of war all the way to its conclusion, thus providing the opportunity to gauge the extent to which policy is shaped, and possibly trapped, by its interplay with hostility and chance. In addition, the wars of former Yugoslavia also offer an intriguing juxtaposition to the new war thesis which identified that conflict as the first new war; and which inadvertently revivified Clausewitzian scholarship. These wars were highlighted by the new war writers as particularly brutal examples of the changing nature of war. More particularly, they were used to explicate the new war idea by rejecting the Clausewitzian Trinity; with a perceived paucity of political rationality being used to demonstrate the obsolescence of Clausewitz's claim that war is a continuation of policy.

Furthermore, the Balkan wars were the first in post-Cold War Europe, and exhibited, at least superficially, many of the characteristics of what has been described as a new type of war. They provided the lexicon of international relations with a new term – 'ethnic cleansing', and marked a turning point in the way in which international society responded to state violations of human rights. This was most clearly demonstrated when the United Nations Security Council unanimously approved Resolution 827, establishing 'an international tribunal for the sole purpose of prosecuting persons responsible for serious violations of international humanitarian law in the territory of former Yugoslavia'.[59] The first of three ad hoc international tribunals, it would shape the way in which war and human rights abuses were understood by the international community – indictments would not be limited to those carrying out crimes and would reach as far as Heads of State, a position which seemed to undermine the very idea of sovereignty in international relations.[60]

---

59   The International Criminal Tribunal for the Former Yugoslavia – UN Security Council Resolution 827. UN Doc. S/RES/827 (1993).

60   The other two tribunals were set up to deal with the wars in Rwanda and Sierra Leone. See: William A. Schabas, *The UN International Criminal Tribunals: The Former Yugoslavia, Rwanda and Sierra Leone* (Cambridge: Cambridge University Press, 2006). A succinct examination of the political motivations that led to the ICTY can be found on pages: 13–24. The most famous of the indictments reached to the very top of the Serbian

For many analysts, it was the wars of former Yugoslavia that projected the idea that the nature of war was undergoing profound change. Of course, in the words of Kaldor, it was 'the archetypal example, the paradigm of the new type of warfare'.[61] As the war in Yugoslavia became headline news during the 1990s, and as stories of atrocities more reminiscent of World War Two began to surface, the new war writers appeared to have the empirical evidence to support their claims. The Balkan wars seem to display all the characteristics of the new war phenomenon – non-state actors, paramilitary groups, ethnic cleansing and genocide are all elements of the wider new war argument. All are found in the wars in former Yugoslavia. For new war theorists, this conflict seems to fit the bill perfectly.

For many analysts these Balkan conflicts represent the negation of Clausewitzian theory altogether. After all, how can the Clausewitzian claim that war is a 'continuation of rational policy' make sense of ethnic cleansing and genocide on the scale experienced in the Balkans in the 1990s? For those who reject Clausewitz's theory, the war's ferocity, acts of genocide and ethnic cleansing all contribute to the opinion that the war was irrational, devoid of political control and driven by cultural and religious hatred. Exhibiting a strange predilection towards warfare as the favoured way of settling disputes, it seemed that the Balkans suffered from some peculiar regional malady. The scale of the fighting, and the manner in which Yugoslavia descended into a maelstrom of war, of claim and counter claim over territory, appears to have been the evidence needed to substantiate the argument that this was a post-modern conflict. Certainly, it was viewed by many as a conflict that lacked the political rationality of past wars.

Underlying this argument is the still popular belief that the people of the region are unusually volatile with a tendency to religious and cultural intolerance. The causes of the conflict were thought to be an outpouring of pent up cultural and ethnic resentment fostered during the years of socialist Yugoslavia – past wrongs incurred over centuries were presented as the primary motivating factors for the outbreak of war. As the war spread from Slovenia, to Croatia and then Bosnia-Herzegovina, this argument developed quickly. That religion and ethnicity seemed central to the causes and conduct of this war was evidence of war's changing character, clear evidence that war was no longer an affair of rational state policy.[62]

---

state with that of Slobodan Milošević. The leaders of the Bosnian Serb army and parliament, Ratko Mladić and Radovan Karadiz were also indicted. See: The International Criminal Tribunal For The Former Yugoslavia – The Prosecutor of the Tribunal against Slobodan Milošević et al. Case – No. IT-01-50-1. and ICTY – The Prosecutor of the Tribunal against Radovan Karadiz and Ratko Mladic. Case – No. IT – 95-5/18.

61   Mary Kaldor, *New and Old Wars*, 31.

62   Alternative explanations now locate 'elite predation' as a major causal motivation for war. See, for example: Gordy, Eric, *The Culture of Power in Serbia: Nationalism and the Destructions of Alternatives* (Pennsylvania: Pennsylvania State University Press, 1999).

When the new war writers analysed the wars in the former Yugoslavia, the Caucasus and throughout Africa during the 1990s, they appeared to have found the verification to substantiate their claim that conflict in the contemporary world was undergoing transformational change. Economic breakdown, the disintegration of states into ethnic or cultural localities vying for security at the expense of former neighbours, and the purposeful targeting of civilians for tactical advantage have all been highlighted to illustrate their point. As the experiences of warfare in the Balkans have helped shape the opinions of many 'new war' proponents, it is appropriate that the same conflict be used to re-evaluate the utility of the theory they chose to repudiate. As the following chapters illustrate, not only does the Trinity better illuminate the complexity of these wars, the interactive formula emphasized by the concept should provide the strategic platform for thinking about today's conflicts. In doing so, the book identifies the continued dominance of policy within the Trinity, it offers a new interpretation of the Trinity and its place in *On War* – a new critical reinterpretation of Clausewitzian theory.

Of course, exploring Clausewitz's model experientially also chimes with his use of empirical evidence – historical experience – as a core element of his own study. As examined in more detail in the following, this is something clearly in keeping with Clausewitz's own belief that theory should be studied, rather than become doctrine on its own. As he reminds readers, it is 'inquiry which is the most essential part of any theory'.[63] From as early as his '1816–1818 Note' he tells us that: 'Analysis and observation, theory and experience must never disdain or exclude each other; on the contrary, they support each other'.[64] Experience and empirical evidence form a key function in his methodology, using it to test the validity of his conceptualizations against the hard facts of the real world. He dedicates the whole of Chapters Five and Six of Book Two to 'Critical Analysis' and 'On Historical Examples' respectively. In other words, in his pursuit of a theory he critically analyses the experiences of the real world. History acts as a source of empirical knowledge; the evidence he needs to test his ideas.[65] He explains: 'Historical examples clarify everything and also provide the best proof in the empirical sciences'.[66]

Although Clausewitz's supporters have gone a long way in demonstrating the timelessness of his core principles within the Trinitarian construct, there remains a requirement to test the validity of the concept against the idiosyncrasies of real war. Despite Clausewitz's own use of historical examples, neither Clausewitz nor his modern supporters explore the model by setting it against experience. However, doing so provides much needed empirical evidence for the strength and weaknesses of the idea, and in the cases in question points to the continued role of politics as a dominant element. As such this book fills a major gap in our

---

63   Clausewitz, *On War*, 162–3.
64   Clausewitz, 'Note 1816–1818', 69.
65   Clausewitz, *On War*, 199–204.
66   Clausewitz, *On War*, 199.

understanding of the Trinitarian idea, offering new insights into the concept and its place in Clausewitzian theory and modern strategic thought.

Structure

It is necessary to point out that the following chapters encompass a broad conceptualization of war. The book explores the phenomenon of war holistically, not as something solely situated on the battlefield.[67] This is actually at odds with Clausewitz's original intention: he intended his treatise as an operational guide to fellow soldiers. Despite conceptualizing his famous model with the inclusion of policy/politics, his opus is chiefly concerned with war in the narrower sense, with what today we would describe as the operational level. W.B. Gallie observes:

> The accepted view of Clausewitz's philosophy of war is that its core lies in his conception of war as the continuation of policy by the addition of other means, or more simply, of war as a political instrument. But this view, although advocated by genuine admirers of Clausewitz, is liable to mislead. It suggests that his real interest, even if it is focused in many of his chapters on specifically military questions, was of a wider kind: it was in politics, and, more particularly, in the relations, tensions and struggles between different political units.[68]

Gallie continues: 'But the simple and obvious truth is that Clausewitz's main interest was in war … *On War* is emphatically about war, and was primarily written for military men. War is its subject, and the different qualities, relations and dependencies which Clausewitz attributes to war in the course of his book are connected by a single aim: to make clear what this terrible and tragic aspect of human life is about and how it operates.'[69] Gallie is undoubtedly correct. However, it was Clausewitz who injected his treatise with the political element, 'war as a continuation of policy'. Politics provided the synthesis to his examination of war – it is the unifying theme that tied his ideas together. Clausewitz's tells us that the Trinitarian model is intended to 'provide a first ray of thought' on a general theory of war. Whether that would have led to a reworking of his entire treatise remains open to debate. Clausewitz was interested in the nature of war as a pre-requisite to discussing strategic theory more narrowly defined – how to defeat ones' enemy. His theory was primarily about strategy, rather than policy. According to Christopher Bassford, Clausewitz's failure to consider the overarching influence of policy/politics in a more detailed way was simply down to the fact that dealing with war was a big enough job in itself. As Bassford puts it: 'One gargantuan topic at a time

---

67   See: Hew Strachan, 'The Lost Meaning of Strategy', *Survival* (Winter 2003).

68   W.B. Gallie, *Philosophers of Peace and War: Kant, Clausewitz, Marx Engels and Tolstoy* (Cambridge: Cambridge University Press, 1978), 61.

69   Gallie, *Philosophers of Peace and War*, 61.

please'.[70] Nonetheless, it was Clausewitz's insight into the political element of war which has had the most lasting effect, and why his ideas remain so vivid today.

Whether Clausewitz would have strengthened his analysis regarding the relationship between war and politics is a point of debate. However, as Peter Paret has suggested, when his opus is studied against his other writings the nexus between politics and war becomes much clearer.[71] As a result we can confidently extrapolate that Clausewitz comprehended war holistically, as a part of political intercourse. In any case, the enduring fascination in Clausewitz has largely been attributed to his insight that war's nature evolves from political beginnings. Furthermore, the modern comprehension of war is far wider than that of Clausewitz's era, and for Clausewitz's theory to retain its canonical status it must continue to inform our understanding of the entire phenomenon. Consequently, as this book is interested chiefly in the nature of war rather than strategy *per se*, it takes a broader assessment of the place of policy, examining war as a phenomenon not restricted to the battlefield, but as something which must include the assessment of the broader political behaviour of those involved in conflict.

While this introductory chapter has situated the book in the context of change in war over two decades, Chapter 1 explores the enduring interest in Clausewitz's Treatise, before then assessing the inspiration and methodological approach behind *On War*. The following chapter and final part of the introductory section (Chapter 2), 'A Tale of Two Trinities', further examines why the Trinity has been misrepresented in much of the literature. It then goes on to examine the Trinity in detail, providing the foundation to understand the dynamic interaction at the heart of the concept. The focus of the chapter is on the centrality of the 'Trinity' in relation to the rest of Clausewitzian theory – a pre-requisite if we are to be able to accurately analyse the importance of each part of the Trinitarian formula.

Chapters 3, 4, and 5, explore the constituent parts of Clausewitz's Trinitarian concept. A great deal of thought has been put into the structure of the study. How do you test a theory that is interactive and which was intended by its inventor to illuminate complexity? Clausewitz's argument was that hostility, chance and policy cannot be accurately measured. Consequently, to provide a better understanding of the strengths and limits of Clausewitz's 'Wondrous Trinity' I have divided the three constituents of the concept into three chapters, each assessing the different elements within the concept. As the Trinity was intended to display the interplay of hostility, chance, and reason (policy), splitting the concept into constituent parts is somewhat unnatural. Nevertheless, stripping it to its bare bones is the best way of assessing whether the forces it claims to represent continue to impact on modern

---

70   Christopher Bassford, 'Tip-Toe through the Trinity, or The Strange Persistence of Trinitarian Warfare', www.Clausewitz.com/CWZHOME/TRINITY/TRINITY8.htm, (2007), 10 (accessed 11 July 2009).

71   Peter Paret, 'Clausewitz's Politics', *Understanding War: Essays on Clausewitz and the History of Military Power* (Princeton: Princeton University Press, 1993), 167–77.

war. Clausewitz also used this method in his pursuit of a general theory, only adding politics as a unifying synthesis to make sense of the array of divergent forces explicit in war. Thus, to assess the interplay of these forces the Trinity must first be divided. Though separated into constituent parts, the book assesses, collectively, whether the interaction of the competing elements of the Trinity can really illuminate war's nature. Although Christopher Bassford's most recent study of the Trinity has reordered the sequence in which the components are listed, this study stays true to the original. Instead of following the original formula, Bassford argues that hostility and reason are competing parts of the human condition and his discussion therefore reflects that fact. Chance, Bassford notes, is 'everything outside our own skulls'; it is presumed to be neutral from the other elements.[72] This approach misinterprets the importance Clausewitz placed on chance being a natural by-product of reciprocity.

Lastly, it is important to reiterate at this point that the case study is not intended as a new history of the wars of former Yugoslavia and as such the case study does not provide a chronological account of Yugoslavia's fragmentation and war. Instead, by using the war as a context, the study explores the relationship between war, hostility, chance and reason (policy). These three chapters provide the case study portion of the book and illuminate the modern utility of Clausewitz's model. Moving from hostility, through chance, the case study is able to better understand the influence of these elements of the triad. The book then explores the ways in which these influences affected policy. As noted at the beginning of this introduction, the book argues for the essential validity of the Trinitarian model. However, in a major departure from recent works it argues that the essential insights of the Trinity should not undermine the central importance of politics in Clausewitz's tome. Indeed, as the case study illustrates, policy remains of critical importance to the model and emphasizing its importance can provide a better platform for understanding war in the modern world.

---

72   Bassford, 'Tip-Toe Through the Trinity', 17.

Chapter 1
Clausewitz's Enduring Legacy: Inspiration, and Methodology

If we are to understand and examine the contemporary merits of his Trinity appropriately, it is first necessary to assess Clausewitz's own unfolding understanding of war. This chapter consequently acts as a reference point for the remainder of the book. It explores the enduring interest in Clausewitz's work, and examines both his inspiration to write *On War*, and the complex methodological underpinnings which resulted in his 'Wondrous Trinity'. As Clausewitz's early death complicates interpretation, before further presenting the Trinity it is important to provide a better appreciation of his inspiration and methodological approach.

On War's Enduring Legacy

Interpretive questions notwithstanding, Clausewitz's masterpiece continues to rank in the top tier in the canon of great books on war.[1] As the American strategist Bernard Brodie famously remarked, it is 'not simply the greatest but the only truly great book on war'. It is a sentiment shared by Colin Gray, for whom *On War* remains 'the Gold Standard of general strategic theory'.[2] Yet despite such evident homage, Clausewitz's ideas have resulted in a great deal of spilt ink over the years. The conjecture which now surrounds the Trinity should not be terribly surprising. It has aroused unending interpretative speculation since its posthumous publication. Of course, that its ideas continue to engender so much controversy is in large part because they are somewhat inaccessible; a consequence of the author's untimely death. As Clausewitz makes clear in a prefatory note written in 1830:

> The manuscript on the conduct of major operations that will be found after my death can, in its present state, be regarded as nothing but a collection of materials from which a theory of war was to have been distilled. I am dissatisfied with

---

1   Bernard Brodie, 'The Continuing Relevance of *On War*', in Clausewitz, *On War*, Michael Howard and Peter Paret transl., 50.
2   Colin S. Gray, *Modern Strategy* (Oxford: Oxford University Press, 1999), 112.

most of it ... I intended to rewrite it entirely and to try and find a solution along
other lines.[3]

Although Brodie considered the ideas within *On War* to be 'generally simple and
for the most part clearly expressed in jargon-free language', this is somewhat
misleading.[4] The ideas in Clausewitz's opus have ensured him a degree of
immortality, but the combination of complex ideas, a paucity of methodological
clarity and the unfinished nature of his work does not facilitate easy reading. As
one Clausewitzian scholar put it, 'the fact that it (*On War*) towers above the rest of
military and naval literature, penetrating into regions no other military thinker has
ever approached, has been the cause of its being misunderstood'.[5] It was a problem
that Clausewitz was only too well aware of. Writing somewhat prophetically in his
note of 1827, he remarked:

> If an early death should terminate my work, what I have written so far would,
> of course, only deserve to be called a shapeless mass of ideas. Being liable to
> endless misinterpretation it would be the target of much half-baked criticism.[6]

And Clausewitz's ideas have suffered from their fair share of criticism and
distortion. In reference to this misrepresentation one thinks immediately of his
alleged complicity as architect of the Great War (1914–18). Derided as the 'Mahdi
of mass and mutual massacre' by Basil Liddell Hart, Clausewitz was accused of
propagating the phenomenon of total war, of which the Great War seemed the
ultimate manifestation.[7] After all, was it not Clausewitz who instructed that war
must reach its 'ideal' type – absolute war? Was it not Clausewitz who preached
about the centrality of battle? In Chapter One, Book One, the only chapter
Clausewitz considered completed, he argues that 'war is an act of force, and there
is no logical limit to the application of that force'.[8] On the face of it, it would
seem his critics have a point. Yet, the problem for Clausewitz's opponents – past
and present – is that extracts such as this only delimit Clausewitz's work when
presented out of context.

---

3   Carl von Clausewitz, 'Note of 1830' (1882) *On War*, transl. by Michael Howard
and Peter Paret (New York: Alfred A. Knopf, Everyman's Library edition, 1993), 79.
4   Brodie, 'The Continuing Relevance of *On War*', 50.
5   Herbert Rosinski, *The German Army*, edited and with an introduction by Gordon
A. Craig (London: Pall Mall P., 1966), 122.
6   Clausewitz, 'Note of 10 July 1827', *On War*, 74.
7   Basil H. Liddell Hart, *The Ghosts of Napoleon* (London: Faber and Faber, 1933),
120. For more on Liddell Hart, see: Brian Bond and Martin Alexander, 'Liddell Hart and
De Gaulle: The Doctrines of Limited Liability and Mobile Defence', in Peter Paret (ed.),
Makers of Modern Strategy (Princeton: Princeton University Press, 1986), 598–623.
8   Clausewitz, *On War*, 85.

Since his death, Clausewitz's ideas have been used to prop up a variety of political and theoretical projects which were patently not the intention of the author, shaped to meet the changing circumstances of each period.[9] Liddell Hart's antipathy towards Clausewitz was linked to a misinterpreted fascination with total war, which is at odds with what Clausewitz was really saying. It has been tacitly inferred that Clausewitz was delineating the centrality of battle so as to expose the frailty of limited means [10] However, Clausewitz's opus is much more than a simple exposé of war's violent proclivities. Explicative rather than purely didactic, it strives to understand war in order to equip soldiers with guidelines, which would serve as the necessary foundation, a basis for action in a labile business requiring intellectual acuity. *On War* was not intended to inculcate soldiers with immutable rules; Clausewitz believed prescription deleterious, criticizing such efforts not least because they are likely to impede action by stifling the intuition of the commander. His real/ideal dichotomy, though the source of a great deal of confusion and rancour, does not propound the mantra of annihilation that many commentators previously assumed. Clausewitz's conception of 'ideal' or 'absolute' war was an abstraction which can never be reached. Its reification is not attributable to Clausewitz, but to later scholars who moulded his ideas to support their own contentions.

As Raymond Aron, one of the twentieth century's great proponents of Clausewitz complained, the problems in drawing out Clausewitz's core message are compounded by the fact that different parts of his tome seem to contradict each other. Aron remarks, 'you can find what you want to find in the Treatise: all that you need is a selection of quotations, supported by personal prejudice'.[11] It is a problem which continues to arouse reproach. As one recent critic has argued, the fact that Clausewitz's ideas can produce infinite interpretations makes it 'of little practical utility'.[12] Another detractor, Bruce Fleming, a Professor of English at the

---

9   Probably the most extreme case was the use of Clausewitz to underpin the message of the Nazi Party in Germany. At a time of national introspection, Nazi ideologues were quick to draw parallels with Clausewitz's own experience. Clausewitz too had suffered the humiliation of defeat and his message of an existential struggle could be unpackaged to prepare Germany for the new struggles of the day and to overcome the shame of defeat in 1918. See: P.M. Baldwin, 'Clausewitz in Nazi Germany', *The Journal of Contemporary History*, 16 (1981). Although Clausewitz's writings were frowned upon by Stalin, Clausewitzian ideas have a long history in Soviet military thought. See: Heuser, Beatrice, *Reading Clausewitz* (London: Pimlico, 2002), 143–50. For a thoughtful exposition of Clausewitz's influence on German Military thought, see: Antulio J. Echevarria II, *After Clausewitz, German Military Thinkers Before the Great War* (University Press of Kansas, 2000).

10   Important works include: Stephen J. Cimbala, *Clausewitz on Escalation; Friction in War and Military Policy* (Westport, Connecticut: Praeger, 2001).

11   Raymon Aron, *Clausewitz, Philosopher of War*, transl. by Christine Booker and Norman Stone (London: Routledge, 1983), 235.

12   Tony Corn, 'Clausewitz in Wonderland', Policy Review – Web Special (September, 2006), http://www.hoover.org/publications/policyreview/4268401.html (accessed 23 October 2013).

United States Naval War College, has even gone as far as to argue that it should be regarded as poetry, 'as an expression of the intrinsic contradictions of the human condition'.[13]

Of course, as Bruce Fleming also points out, the very fact that its ideas are open to interpretation is one of the reasons why *On War* continues to sustain such enduring interest; provoking varied conclusions for the people who read it.[14] It may generate dissonant views, but the fact that people continue to engage with the subject is in itself no bad thing. Being open to interpretation has positive as well as negative effects. Antulio Echevarria, for instance, has described it as an 'unfinished symphony', which he suggests should remain unfinished. As he puts it, 'While the temptation to finish such works may be great, the results are rarely satisfying. We always seem to be left with the nagging sense that the master would have done it differently'.[15]

He may well be right. *On War's* unfinished nature is partly responsible for it having retained such salience in a subject that, technologically and characteristically at least, has undergone such profound changes since the events which first inspired Clausewitz to write about it – the upheavals of the French Revolution and the ensuing Napoleonic wars. That his ideas remain so redolent today is partly attributable to his vivid insights regarding war's nature and the nexus between war, policy and strategy. Unlike those of his contemporaries, Clausewitz's arguments and ideas continue to inform. If approached carefully and thoroughly, *On War's* content has the power to influence and educate the modern reader; the Trinity is a case in point. On the down side, while Clausewitz's explication of war's nature stands in contradistinction to the theories of his contemporaries, the explanatory purpose of this is confused because the remainder of his treatise has a normative function, advising soldiers and strategists of the best ways to overcome war's unpredictable tendencies. The explicative side of war sits rather awkwardly with Clausewitz's instructions to his readers.[16] It would be useful to explain specifically which parts of the book comprise Clausewitz's 'explication of war's nature'.

Nonetheless, this process of stripping bare the complexity of Clausewitz's work has been particularly helped by *On War's* current popularity; although the profusion of critical texts can engender problems of their own. The development of Clausewitzian thought in English was propelled in 1976 following the publication of the Michael Howard and Peter Paret translation of *On War*, which has opened Clausewitz's ideas to a much wider audience. That Paret also published his landmark biography of Clausewitz's life and politics, *Clausewitz and the State*, in

---

13   Bruce Fleming, 'Can Reading Clausewitz Save Us from Future Mistakes?', *Parameters* (Spring 2004), 76.

14   Ibid.

15   Antulio J. Echevarria II, *Clausewitz and Contemporary War* (Oxford: Oxford University Press, 2007), 7.

16   See: Jon Tetsuro Sumida, *Decoding Clausewitz: A New Approach to On War* (Kansas: University Press of Kansas, 2008).

the same year provided yet more impetus to engage with Clausewitz's previously esoteric musings.[17] Along with Raymond Aron's *Clausewitz: Philosopher of War* – a major critical enquiry into Clausewitz's ideas, interest and speculation of how his ideas transported to contemporary situations grew. At a time when the world was locked into the Cold War, Clausewitz's ideas regarding the relationship between policy and war, as articulated in the Howard and Paret translation and the accompanying works by Paret and Aron, were viewed as essential to understanding and planning for war in the nuclear age. On the back of America's humiliating defeat in Vietnam, the Clausewitzian formula – war as a continuation of politics – took on even greater significance. The US Military gradually adopted Clausewitzian theory as a key component in the preparation of military doctrine.[18] If the US military has continued to eschew Clausewitzian tenets, it is not for want of engagement by Clausewitzian scholars.

As the corpus of literature has blossomed, so too have essential examinations of how Clausewitz's ideas have been transported and recalibrated from one period to the next. This explosion of critical thinking owes a huge debt to Howard and Paret. Although any translation is also a work of interpretation and can thus be challenged, they have helped underpin the study of war and strategy with the theoretical knowledge required for serious reflection.

Notwithstanding the merits of the Howard/Paret translation, their approach has been disputed. Implicit in their translation, and explicit in Paret's *Clausewitz and the State*, is the contention that the political-war nexus is clearly definable throughout the entirety of *On War* despite its fragmentary and incomplete nature. This is challenged chiefly by Azar Gat, who contends that Clausewitz only realized the significance of the political nexus as late as 1827, causing him to revise his whole thesis. The debate has been further complicated by another school of thought claiming that *On War* presents a comprehension of war's existential side. The chief proponents of this claim are Andreas Herberg-Rothe and his former supervisor Herfried Münkler. Yet, whether this has resulted in 'turmoil' within Clausewitz studies as Hew Strachan contends, is open to debate.[19] In other disciplines such a dispute would be construed as being healthy,

---

17   Peter Paret, *Clausewitz and the State. The Man, His Theories, and His Times*, 2007 edition (Princeton: Princeton University Press, 2007).

18   Stuart Kinross, 'War is an Instrument of Policy': The Influence of Clausewitz upon American Strategic Thought and Practice from the Vietnam War to the Gulf War' (Unpublished PhD, University of Aberdeen, 2001). Now published as Stuart Kinross, *Clausewitz and America: Strategic Thought and Practice from Vietnam to Iraq* (London: Routledge, 2005). For an examination of Clausewitz's reception in British and US military thought prior to 1945, see: Christopher Bassford, *Clausewitz in English – The Reception of Clausewitz in Britain and America, 1815–1945* (New York: Oxford University Press, 1994).

19   Hew Strachan, 'On Clausewitz: A Study of Military and Political Ideas,' A Review Essay. *War in History*, 14(2) (2007), 280–32.

preventing scholarship from becoming staid. The very fact that so many scholars are embracing the seriousness of the subject is no bad thing.

Clausewitz: Inspiration and Methodology

While Clausewitz did not complete his intended revisions to *On War* before his death, his hope that his treatise contained 'the basic ideas that might bring about a revolution in the theory of war' was certainly granted.[20] His opus may remain unfinished, but this has done little to limit its importance; and *On War* has certainly been important. However, its significance does not rest with the fact that it is open to debate. Its real strength, and the reason for its adaptability and durability, stems from Clausewitz's own motivation to write, the methodological approach he employed, and, most importantly, the findings and concepts that originated from his study.

The impetus to undertake a study of such magnitude originated from Clausewitz's personal experience as a soldier in the Napoleonic Wars, particularly in the battles of Jena-Auerstadt (1806). He was a witness to the annihilation of the Prussian army by the *Grande Armée* of Napoleon at Auerstadt. After crushing the Third Coalition (Austria, Portugal, and Russia) the previous year, the ruin of the Prussian army cemented Napoleon's position as master of Europe. Clausewitz was taken prisoner as an aide to Prince August of Prussia and held in France until 1807, and his experience of French internment invoked a visceral dislike for all things French and an equally profound belief in the superiority of the Germanic peoples that would last his entire life. Indeed, in an early essay comparing the French and German characters he remarked, 'it (the French character) is inconsistent and not very deep'. Such character traits sit in sharp contrast to those of the Germans, which, he opines 'are almost completely the opposite'.[21] Yet despite these frustrations, Clausewitz's captivity provided ample time to reflect upon the 'new' dynamic of war which had carried the Napoleonic armies to sweeping victories across Europe. Undeterred by his dislike of French custom, he set himself the task of understanding the new realities of war, so clearly demonstrated by Napoleon. He spent the remainder of his life studying war in pursuit of a general theory which could explain the nature of war, and which could prepare Prussia for more successful times ahead.

---

20   Carl von Clausewitz, 'Note of 1827' (1882) *On War*, transl. by Michael Howard and Peter Paret (New York: Alfred A. Knopf, Everyman's Library edition, 1993), 78.

21   Carl von Clausewitz, 'The Germans and The French' (1807), in *Carl Von Clausewitz: Historical and Political Writings*, transl. by Peter Paret and Daniel Moran (Princeton: Princeton University Press, 1992), 255, 256. See also: Peter Paret, *Clausewitz and the State. The Man, His Theories, and His Times*, 2007 edition (Princeton: Princeton University Press, 2007), 132–5.

Once released from French custody, Clausewitz immediately rejoined the army. Returning to Prussia he became a key figure in the coterie of influential young officers entrusted with the reform of the Prussian army by Gerhard von Scharnhorst – one of the key figures in Clausewitz's intellectual life.[22] He thus played a key role in the realization of the politically sensitive 'Military Reorganization Commission' which had been tasked by Scharnhorst to review Prussian military practice in the aftermath of Prussia's humiliating defeat at the hands of Napoleon.[23] After graduating at the top of his class at the new staff college in Berlin, Scharnhorst's patronage enabled Clausewitz to take up positions of importance within the reform movement. This propelled his career within the army. However, his role as a 'reformer' also pitted him against the conservatism of the royal government, which grew to view him as a political recalcitrant who disseminated ideas on military reform which clashed with traditional norms. It was a position which would prove injurious to his future career; most notably when he was denied the opportunity to embark upon a diplomatic career as Prussian Minister to the court of St James.[24]

There is scholarly consensus that Clausewitz was no revolutionary, yet he was indignant and often vitriolic in his condemnation of the antiquated system of military culture and patronage that had proved so disastrous in Prussia's miserable capitulation to the French.[25] In Clausewitz's view, one of the reasons for Napoleon's early victories was the opportunity for gifted men to take positions of importance within the army, a situation stemming from the French Revolution. Napoleon's army was not corralled by traditional social custom and patronage, which so stifled the armies of the *ancien régime*. In Clausewitz's view, Prussia's defeat at Jena-Auerstadt stemmed not from an exiguous appreciation of the new methods of war, but a failure of the state's military leadership to lead or inspire – a political rather than purely military cause. Peter Paret writes:

> The defeat of Prussia in 1806 confirmed Clausewitz in his view that war could not be considered in isolation, as an essentially military act. It was obvious to him that politics of the previous decade had largely decided the issue before

---

22   For more on Scharnhorst's own rise to prominence, and career, see: Paret, *Clausewitz and the State*, 56–77. See also: Clausewitz, 'On the Life and Character of Scharnhorst' (1817), in *Carl Von Clausewitz: Historical and Political Writings*, transl. by Peter Paret and Daniel Moran (Princeton: Princeton University Press, 1992), 85–109.

23   Another examination of the German reform movement can be found in: Azar Gat, *A History of Military Thought: From the Enlightenment to the Cold War* (Oxford: Oxford University Press, 2001), 139–57.

24   Carl Von Clausewitz, '"A Proposition Not a Solution" – Clausewitz's Attempt to Become Prussian Minister at the Court of St. James', in Peter Paret (ed.), *Understanding War: Essays on Clausewitz and the History of Military Power* (Princeton; Princeton University Press, 1993), 178–90.

25   Carl Von Clausewitz, 'Observations on Prussia in Her Great Catastrophe' (1823–25), in *Carl Von Clausewitz: Historical and Political Writings*, 30–85.

the fighting began, while social conditions of long standing in the Prussian monarchy had created military institutions and attitudes that proved hopeless against an opponent who was numerically superior and attuned to new forms of fighting.[26]

These views undoubtedly created dissonance in the upper echelons of the army, and his conviction that the army should abolish the traditional habit of awarding positions of power reflecting social rank was viewed as dangerous, and possibly even subversive. Although he succeeded in reforming the army to incorporate the *Landwehr* system of territorial militias, which helped Prussia defeat the Napoleonic armies after 1813, when the crisis had passed and when the state felt more secure again, the work of the reformers was reversed.[27] The fact that Clausewitz came from a different social stratum to many of his contemporaries only added to doubts about his political sensibilities. As a consequence he was viewed with suspicion outside the immediate environs of the reform movement.

Throughout his imprisonment in France and throughout his association with the reform movement, Clausewitz's personal development, and his unfolding understanding of Prussian/German nationalism, pitted him against the state he sought to serve. Clausewitz supported the status quo, but he was tendentious in his repudiation of existing military practice and his historical and political writings clashed with conservative values. His support for the King was real enough, but he expected the Crown to fight to the death in the name of Prussian nationalism. In defeat, especially after Jena-Auerstadt, he called for a renewal of hostilities as a means of revivifying the nation. In Clausewitz's 'Political Declaration', written in protest against the French suzerainty of Prussia, which had culminated in the French order to support its planned invasion of Russia in 1812, he proposes a 'peoples' war' to defeat the French, calling for their 'alliance' to be immediately revoked.[28]

Clearly, in his mature thinking Clausewitz is intimately associated with the idea that war is instrumental, a 'continuation of policy'. Yet, this existentialist side never left him and it is an element in his thinking which was never fully resolved in the years preceding his death. Indeed, we can extrapolate from the available evidence that in certain circumstances such existentialism is wholly consistent with his other ideas.[29] Even when war is employed as a vain expression of the nation, it retains its instrumentality. He argues in *On War*:

---

26   Peter Paret, 'The Genesis of *On War*', *in* Clausewitz, *On War*, 13.

27   See: Clausewitz, 'On the Political Advantages and Disadvantages of the Prussian Landwehr' (1819), in *Carl Von Clausewitz: Historical and Political Writings*, 329–35.

28   Clausewitz, 'Political Declaration', *Historical and Political Writings*, 285–303.

29   Hew Strachan, 'Clausewitz and the Dialectics of War', in Hew Strachan and Andreas Herberg-Rothe (eds), *Clausewitz in the Twenty-First Century* (Oxford: Oxford University Press), 36.

There will always be time enough to die; like a drowning man who will clutch instinctively at a straw, it is the natural law of the moral world that a nation that finds itself on the brink of the abyss will try to save itself by any means.

No matter how small and weak a state may be in comparison with its enemy, it must not forgo these last efforts, or one would conclude that its soul is dead.[30]

When Napoleon invaded Russia in the campaign of 1812 ordering a Prussian contingent to join his ill-fated invasion, Clausewitz refused to cosset royal sensibilities and campaigned for Prussian dissociation from Napoleon's France. In Clausewitz's view, Prussia should choose to fight a 'people's war' against the French.[31] When this was refused, he took a commission in the Russian army rather than act as an adjunct in French imperial adventures. As the following quote reveals, his visceral attachment to the German nation was clear:

I would use lashes of the whip to arouse the animal from lethargy, so that the chain it has allowed to be placed upon it in such a cowardly and timid way would be shattered. I would set free in Germany a spirit that would act as an antidote, using its destructive force to eradicate the scourge that threatens to cause decay of the entire spirit of the nation.[32]

Instrumental in brokering the Convention of Tauroggen (1812), which ended Prussian military support for Napoleon in his invasion of Russia, he was never forgiven for his decision to enter Russian service. For the Prussian crown this was viewed as betrayal, another example of Clausewitz's dangerous revolutionary sentiment.[33] Despite such setbacks, Scharnhorst's patronage allowed him the opportunity to think and write about war, and his time at the heart of the reform movement provided the perfect platform from which to theorize and write about the subject that dominated his life. Retrospectively, the very fact that his career never took the active path which Clausewitz had once hoped for provided him with the opportunity to write *On War*.

---

30   Clausewitz, *On War*, 583.

31   W.B. Gallie, *Philosophers of Peace and War: Kant, Clausewitz, Marx Engels and Tolstoy* (Cambridge: Cambridge University Press, 1978), 39.

32   Clausewitz, *Schriften, Aufsätze, Studien Briefe*, edited by Werner Hahlweg, 2 vols, Gottingen 1966 and 1990, 686–7 – quote found in Herberg-Rothe, *Clausewitz's Puzzle*, 26.

33   Clausewitz was not fully readmitted to the army until 1815, and thus he was a bystander to the Prussian campaigns against France during 1813–14. As Gallie (1978: 39) remarks, 'His plans for a popular militia (*landstrum*) to support the regular army against the French were dismissed as a recipe for revolution.

A Clausewitzian Methodology? An Empirical Approach

Although one would expect a major work such as *On War* to have a clear methodological structure, the reader is left disappointed. The reason is undoubtedly due to the unfinished nature of the work, and one would hope that Clausewitz would have provided a clearer picture of what he was doing, and how he would go about it, had he lived to finish his treatise. As a means of better understanding the Trinity, the present section explores his methodology.

Empowered by the support of Scharnhorst, Clausewitz rejected what he considered to be the staid classifications advanced by other military theorists. Using his own experiences as a foundation for study, he also gleaned insights through collaboration with his peers at the war college and on the staff of Scharnhorst's ministry, where he worked closely with former classmates, especially Karl von Tiedemann and August Gneisenau – both of whom owed their positions to Scharnhorst's benefaction.[34] Although contemporaries in the reform movement shared Clausewitz's gusto for change, it was Clausewitz's ability to think holistically and abstractly about war, and then articulate his thought, which allowed his theory to surpass the ideas generated by his contemporaries.[35] It is in Clausewitz's work that the nexus between the social upheavals of the French Revolution and the resulting revolutionary changes in war is fully explored. His experiences during the Napoleonic wars demonstrated the power of emotion and passion, of nationalism, but also of chance and friction, features of war that had been neglected by his contemporaries.

It was the realization that the power unleashed by the French Revolution could so massively influence war – so aptly demonstrated by Napoleon – which focused Clausewitz's early thoughts, and which directed his initial study. *On War* is replete with references to battle and power, a reflection of his early understanding and awe of Napoleonic warfare. As he argues, 'Force – that is, physical force, for moral force has no existence save as expressed in the state and the law – is thus the means of war; to impose our will on the enemy is its object'. He continues: 'To secure that object we must render the enemy powerless; and that, in theory, is the true aim of warfare'.[36] Force is emphasized again when he notes that: 'Fighting is the central military act; all other activities merely support it'. If there is no fighting, then, there is no war: 'Total non-resistance would not be war at all'.[37] In yet another passage he reminds readers that: 'The fighting forces must be destroyed: that is, they must be *put in such a condition that they can no longer carry on the fight*'.[38] 'It follows, then, that to overcome the enemy, or disarm him – call it what you

---

34   Paret, *Clausewitz and the State*, 141.
35   Beatrice Heuser, *Reading Clausewitz* (London: Pimlico, 2002), 10.
36   Clausewitz, *On War*, 83.
37   Ibid., 86.
38   Ibid., 102.

will – must always be the aim of warfare'. Clausewitz is obdurate, unambiguously emphasizing the centrality of battle:

> Kind-hearted people might of course think there was some ingenious way to disarm or defeat an enemy without too much bloodshed, and might imagine this is the true goal of the art of war. Pleasant as it sounds, it is a fallacy that must be exposed.[39]

With the risk of stating the obvious, for Clausewitz the military clash is central to his understanding of war. This, he claims, is 'for all major and minor operations in war what cash payment is in commerce'.[40]

Yet despite these many references to battle, *On War* is more than an exposition of war's inherent violence. An example can be found in Clausewitz's description of genius. For Clausewitz, the *coup d'oeil*, the emphasis on genius and initiative, is as critical to his understanding of war as military power. Only the initiative of a genius can recognize and overcome the myriad cases of emotion, chance and friction that make war so unpredictable. Clausewitz may have been in awe of the force of Napoleon's advance and the power which had been unleashed through the *levée en masse*, but even when reflecting on the disastrous battles of Jena-Auerstadt, Clausewitz insisted more was achievable by the Prussian army. There can be little doubt as to the impact of the social forces encapsulated in the 'new' French way of war. However equally importantly for Clausewitz, much of the French success could be attributed to Napoleon himself, the 'god of war'. He dedicated Chapter Three, Book One 'On Military Genius' to this topic.[41] Sheer power may be a crucial element, but it was certainly not the only aspect of warfare. He reminds his readers:

> What we must do is survey all those gifts of mind and temperament that in combination bear on military activity. These, taken together, constitute *the essence of military genius*.[42]

While his contemporaries preached the centrality of scientific principles as the key strategic accomplishment, *On War* has a different purpose. Clausewitz's opus is descriptive, an explanatory text which students can use as a guide to their own unfolding comprehension of war's inherent complexity. This partly reflects Clausewitz's comprehension of genius; for him it is an un-teachable gift. Attempting to inculcate students with prescriptive military formulas would stunt the very traits of intuition and freedom of mind that Clausewitz believed great commanders must display. As a consequence, *On War* rejects much of the

---

39   Ibid., 83–4.
40   Ibid., 111.
41   Ibid., 115–31.
42   Ibid., 115.

dogmatism of his generation, shunning the proscriptive markers that many believe he represents. He firmly rejected earlier theories on war, most notably the works of his contemporaries, Antoine Henri de Jomini and Dietrich von *Bülow*, both of whom he derided because they sought exact and immutable rules capable of prescribing blueprints for military success.[43] Rather disdainfully, Clausewitz comments on his own decision to write his tome that:

> Perhaps it would not be impossible to write a systematic theory of war, full of intelligence and substance; but the theories we presently possess are very different. Quite apart from their unscientific spirit, they try so hard to make their systems coherent and complete that they are stuffed with commonplaces, truisms, and nonsense of every kind.[44]

Although *On War* has regularly been misunderstood – viewed and condemned for its ostensible rationalist message regarding the relationship between politics and war – Clausewitz recoiled from the prescriptions of his contemporary theorists. In reflection, he comments that: 'Efforts were therefore made to equip the conduct of war with principles, rules, or even systems ... but this failed to take adequate account of the endless complexities involved'.[45] Clausewitz reminds readers that a theory of war should provide the commander with a platform from which to engage, not by rote, but rather by intuition and acuity founded upon a firm comprehension of war's inherent complexity. As he comments in Book Two, Chapter Two, under the heading 'A Positive Doctrine Is Unattainable',

> Given the nature of the subject, we must remind ourselves that it is simply not possible to construct a model for the art of war that can serve as a scaffolding on which the commander can rely for support at any time.[46]

Indeed, in the section of the same chapter headed, 'Theory should be study, Not Doctrine', the purpose of theory to the commander, Clausewitz explains, will be to 'light his way, ease his progress, train his judgment, and help him avoid pitfalls'.[47] While it must educate the commander in his self education, it is not meant to 'accompany him to the battlefield'.[48] Theory, Clausewitz explains,

---

43 See: Jomini, Antoine Henri de (1862), *The Art of War*, edited with an introduction by Charles Messenger (London: Greenhill Books, 1992). For an excellent discussion of the limitations of Bulow's work in the context of a changing military environment, see: R.R. Palmer, 'Frederick the Great, Guibert, and Bulow: From Dynastic to National War', in Paret, *Makers of Modern Strategy*, 91–119.

44 Clausewitz, 'Note Between 1816–1818', *On War*, 70.

45 Clausewitz, *On War*, 134.

46 Ibid., 140.

47 Ibid., 141.

48 Ibid.

'should not be a 'manual for action', but rather a guide to the complexities and peculiarities of context.[49]

The focus of his work is on explaining and debating war's salient features. What are the features of war that make it so fluid and unpredictable? If we can illuminate these core elements can we then propose guidelines which may help us 'compel our enemy to do our will'? How should the commander balance the age old problem of successfully relating ends with the available means? Clausewitz's tome illustrates the complexities of war in order to provide students with the insights which may enable them to overcome complications when experiencing war's unpredictability first hand.[50]

Unfortunately, from the armed forces of Clausewitz's Prussia, to students studying his ideas in modern staff colleges – supposedly his core audience – the language, reasoning and concepts which have proven so durable are also difficult to grasp unless one takes the time to study them properly; a criticism also highlighted by Clausewitz's supporters.[51] In an age where time is such an expensive commodity, Clausewitz's lengthy and often laborious study is too often misread; or worse, (a recurrent complaint) not read at all. As he feared, it has been 'liable to endless misinterpretation ... the target of much half-baked criticism'.[52] Many commentators have simply plucked what they needed to shore up their own ideas. With persistent frequency, his ideas are attacked without a clear comprehension of what exactly these ideas are. As highlighted in the introductory chapter, the very idea that Clausewitzian theory should be obsolete in an era of 'new wars' stemmed from a misrepresentative portrayal of Clausewitz's Trinitarian concept.

Although his work is incomplete, Clausewitz did provide instructions with which we can make judgements about how to proceed. In his prefatory 'note of 1827' he remarks: 'I regard the first six books ... merely as a rather formless mass that must be thoroughly reworked once more.'[53] Even by the time of his note of 1830, Clausewitz still thought his work represented 'nothing but a collection of materials'.[54] However, in the same note he also proclaims that: 'The first chapter of Book One alone I regard as finished. It will at least serve the whole by indicating the direction I meant to follow everywhere'.[55] We thus have a platform from which to assess his ideas. The one stumbling block arising from this relates to the fact that it is in this chapter that Clausewitz articulates his belief that 'war can be of two

---

49   Ibid.

50   This method reflects Clausewitz's ideas about education, which are expressed in his letters to the philosopher and writer Johann Gottlieb Fichte, in particular regarding the teaching techniques pioneered by Pestalozzi. See: Paret, *Clausewitz and the State*, 169–208.

51   Christopher Bassford, *Clausewitz and His Works*. www.clausewitz.com?CWZ HOME?CWZSMM?CWORKHOL.htm, 7 (accessed 14 September 2010).

52   Clausewitz, 'Note of 10 July 1827', *On War*, 74.

53   Clausewitz, 'Note of 1827', 77.

54   Clausewitz, 'Note of 1830', 79.

55   Ibid.

kinds'; an idea that first appears as late as his note of 1827. This has engendered a great deal of speculation. Some scholars argue that *On War* should be thought of as two separate books. For others, the proposition that war could be of two types as highlighted in his note of 1827 signifies a crisis point – in Azar Gat's opinion, shattering Clausewitz's conception of war. For others, such as Peter Paret or Daniel Moran, it was more of an 'epiphany'.[56]

With this in mind, how should the modern reader interpret this change of tack highlighted in his note of 1827? What does it tell us about Clausewitz's methodological approach? More importantly still, how does this approach illuminate his Trinity? One consequence of the 1827 note is the suggestion that *On War* should be approached as two separate books. Book One, and to a lesser extent Books Seven and Eight, are thought to represent the mature Clausewitz with his emphasis on politics. The remainder of his work reflects his earlier ideas, inspired by the irresistible force of Napoleonic war; most notably the search for the decisive battle. Beatrice Heuser has described this duality as Clausewitz the *realist* and Clausewitz the *idealist*. This is an attractive way of thinking about his uncompleted thesis, but there is a counterargument that such meddling only helps to distort Clausewitz's true meaning. Taking a different view, Echevarria argues that:

> The problem with this approach is that it assumes Clausewitz's later ideas should take precedence over his earlier ones, that the New Testament should replace the Old one, and that the views of Clausewitz the realist should supersede those of Clausewitz the idealist. However, this assumption is not supported by what he actually said in the note of 1827, or by the revised portions of *On War*.[57]

Echevarria's point is important. Although there has been a tendency to value Clausewitz's later ideas over his earlier ones, these earlier ideas form the basis of his mature thought. Clausewitz the *realist*, and especially his argument about the primacy of politics, infused his theory with a unifying theme; a coherence and universality at variance with the ideas of his contemporaries. Yet, as Echevarria comments, it did so 'without ever diminishing the significance he had already attributed to war's means'.[58] As Christopher Bassford remarks, failure to approach *On War* as one book 'leads to serious misunderstandings of Clausewitz's arguments, for it is precisely through Books Two to Six that he works out the practical implications of his ideas.'[59] Clausewitz may have noted that his 'revision

---

56   See Paret, *Clausewitz and the State*, 147–68; and Daniel Moran, 'The Instrument: Clausewitz on Aims and Objectives in War', in *Clausewitz in the Twenty-First Century*, 93.

57   Antulio J. Echevarria II, *Clausewitz and Contemporary War* (Oxford: Oxford University Press, 2007), 5. This is a view shared by Christopher Bassford, see: Bassford, *Clausewitz and His Works*, 7.

58   Echevarria, *Clausewitz and Contemporary War*, 5.

59   Bassford, *Clausewitz and His Works*, 7.

will also rid the first six books of a good deal of superfluous material', but he also stated that it would 'fill in various gaps, large and small, and make a number of generalities more precise in thought and form'.[60] What is clear is that his existing ideas formed the basis for his revision. This debate has been given extra fuel by the very convincing assertion by Azar Gat that there are real problems over the dates of the prefatory notes. It is Gat's contention that the 1830 note was actually written before the 1827 note.[61] The consequence of this is that Clausewitz's tome is probably more complete than given credit for. It is therefore prudent to approach his book as a whole, rather than break it down into 'old' and 'new' testaments.

In its own right, Clausewitz's early work is important for its insights into war's reciprocal and escalatory tendencies, and acts as a didactic tool which he uses to evolve and mature his own comprehension of war. This was an important part of the process that led him to his core arguments, not least the Trinity. There are strong arguments for approaching *On War* as one work. On this matter, Clausewitz states that:

> I believe an unprejudiced reader in search of truth and understanding will recognize the fact that the first six books, for all their imperfection of form, contain the fruit of years of reflection on war and diligent study of it. He may even find they contain the basic ideas that might bring about a revolution in the theory of war.[62]

He may not have liked his findings, but Clausewitz's evolutionary, unfolding appreciation of war reflects his search for an objective knowledge and his use of empirical evidence as a means of reaching it. As Strachan points out, Clausewitz's aim was to 'achieve understanding through debate, through point and counter-point'.[63] Whether Clausewitz's prefatory notes truly represent a crisis point or merely another insight into his unfolding comprehension of war is open to debate. Crisis or not, his realization that war was of two kinds and, ultimately, that it is governed, albeit haphazardly, by politics, reflects Clausewitz's search for an objective understanding of war. Gat may be correct that this revelation 'shattered' his conception of theory up until that point, but it enabled Clausewitz to present war holistically, allowing him to develop exactly what he set out to do: provide a general theory of war. Clausewitz's ability to reflect on the changing fortunes of Napoleon shaped his mature thought and ultimately brought him to think of war as a Trinity, the focus of this book.

60　Clausewitz, 'Note of 1827', 77.
61　Gat, *A History of Military Thought*, 257–65.
62　Clausewitz, 'Note of 1827', 78.
63　Strachan, 'Clausewitz and the Dialectics of War', 37.

Historical Enquiry

As the earlier passages in *On War* highlight, Clausewitz's early thoughts revolved around the destructive forces which drove Napoleon's armies. This was the key to understanding modern war, and Clausewitz sought to articulate it. In the note of 1827, however, Clausewitz explains that he needs to revise his thesis. Although aspects of the Napoleonic wars suggest brute force, Clausewitz concludes that 'War can be of two kinds, in the sense that either the objective is to *overthrow the enemy* – to render him politically helpless or military impotent ... or *merely to occupy some of his frontier-districts* so that we can annex them or use them for bargaining at the peace negotiations'. This was interrelated to a consequent observation, 'that *war is nothing but the continuation of policy by other means*'.[64] Gat identifies Book Seven as the catalyst for change. This is the book which examined defence and in which Clausewitz proclaims that defence is the stronger form of war, offense the weaker. In the writing-up process, Gat argues, Clausewitz realized that his earlier arguments revolving around power and offensive military engagements were on shaky ground. Clausewitz laments:

> One might wonder whether there is any truth at all in our concept of the absolute character of war were it not for the fact that with our own eyes we have seen warfare achieve this state of absolute perfection ... Are we to take this as the standard, and judge all wars by it, however much they may diverge? ... But in that case what are we to say about all the wars that have been fought since the days of Alexander – excepting certain Roman campaigns – down to Bonaparte? ... We would be bound to say ... that our theory, though strictly logical, would not apply to reality.[65]

According to this view, Clausewitz had thought of war in absolutist form, only to be confronted by reality when he tested his idea against historical enquiry. Gat's position has some merit; however, as Herberg-Rothe has revealed, Clausewitz's use of historical evidence as an arbiter of theory reflects his own experience, and by his explorations of other forms of war, it is likely that he had begun to form an alternative perspective on war prior to 1827. In fact, Clausewitz's other historical and political writings suggest that he was aware much earlier that war was part of a wider political sphere. In his history, *The Principles of War* (1812) and *The Campaign of 1812 in Russia* we can already see him work through many of the problems and experiences that would later become central features of *On War*.[66] What cannot be in doubt is that these changes radically altered the way he understood war. Clausewitz's own unfolding understanding

---

64   Clausewitz, 'Note of 1827', 77.

65   Clausewitz, *On War*, 701.

66   Carl von Clausewitz, *The Campaign of 1812 in Russia*, foreword by Michael Howard (Cambridge, MA: Da Capo Press Inc., 1995).

of war is linked not only to the opening years of Napoleonic warfare, the years when Napoleon's armies seemed imbued with invincibility, but equally, and perhaps more dramatically, it is also linked to Napoleon's disastrous retreat from Moscow, and his eventual defeat at Waterloo.[67] Serving as an officer in the Russian army, Clausewitz witnessed Napoleon's catastrophic retreat and the chaos and death served on the French at the 'Holocaust at the Berezina'.[68] His experience of the Napoleonic army's reversal of fortune in Russia and of its eventual defeat at Waterloo – which he partly attributed to an unfavourable domestic political situation in France – demonstrated to him the range of factors which could impede victory. Even Napoleon's genius, his *coup d'oeil*, could be overcome by factors such as territory, friction, and politics. The experience of Napoleon's ruinous adventure in Russia and his defeat at Waterloo highlighted two major problems in finding the correct balance between the purpose of war and the means employed. Firstly, the effects of friction caused by war's reciprocal nature and other myriad neutral factors inhibited Napoleon, sapping his power as a result. Secondly, political interference stemming from the domestic political environment in France impacted on Napoleon's ability to fight with overwhelming strength in Russia and at Waterloo. As Clausewitz realized, the very reason for victory in one situation could be the cause of defeat in another. While Napoleon's recklessness and daring had won him celebrated victories throughout Europe, this characteristic also led him to risk all in his Russian misadventure.

This may indeed have constituted an emergency in Clausewitz's theoretical investigation, but his attempt to rediscover universal truths allowed him to widen his theoretical approach, with empirical evidence and historical enquiry forming key components of his methodological approach. Paret has remarked that Clausewitz's use of historical evidence 'suggests a writer who in historiographical terms is a transitional figure: a rigorous thinker who has left past conceptions behind but has not acquired the new methodological tools that are being developed; an amateur scholar, not an academic, untouched by the nascent professionalism of the discipline of history'.[69] His use of historical evidence is a product of his close relationship with his mentor Scharnhorst and 'cannot be exaggerated'; it enables Clausewitz to elucidate a theory of war not restricted

---

67   Andreas Herberg-Rothe, *Clausewitz's Puzzle* (Oxford: Oxford University Press, 2007), 15–38.

68   In his book on the 1812 Napoleonic campaign in Russia, he remarks about the ferocity and the appalling suffering endured by the French army, even indicating that he would never forget the scene. See: Carl von Clausewitz, *The Campaign of 1812 in Russia*. For a wonderfully vivid account of the battle, see: Paul Britten Austin, *1812: Napoleon's invasion of Russia*, forward by David G. Chandler (London: Greenhill Books, 2000).

69   Peter Paret, 'Clausewitz as Historian', *Understanding War: Essays on Clausewitz and the History of Military Power* (Princeton: Princeton University Press, 1993), 142.

by the parameters of his own era.[70] As highlighted in the subsequent section, Clausewitz was clearly cognizant with theoretical and philosophical methods, yet his exposure to war, and his use of history as an arbiter of his work, is a core element in his theory of war; and a significant part of the process which brings him to his Trinitarian formula. Though Clausewitz dedicates Chapter Six, Book Two 'On Historical Examples' to this element of enquiry, he also notes the difficulty of recreating historical cases. To balance these difficulties, Clausewitz employs a range of techniques drawn from his theoretical and methodological influences – not least his opinion in the preceding chapter of the same book that events must be properly studied, with due care to trace cause and effect as far possible.

It is likely that the crisis in Clausewitz's thinking, great enough for him to begin revision of his entire tome in 1827, was produced by his personal experiences as a serving soldier. However, these same experiences were augmented by historical evaluations which add width and depth to Clausewitz's ideas, in turn refocusing his study away from his earlier preoccupation with power. The Napoleonic armies had demonstrated the effect of brute force, but even when war seemed to reach its ideal type, as Clausewitz believed it had with Napoleon's armies, it was still susceptible to a range of other factors, not least politics and friction. Political interference – the political aims of initiating conflict, and the friction encountered in fighting such difficult campaigns – restricted war's escalatory proclivities. War did not necessarily escalate as he had earlier assumed; escalation was determined by external political factors, as well as neutral factors like friction. As Clausewitz famously explains, 'Everything in war is very simple, but the simplest thing is difficult. The difficulties accumulate and end by producing a kind of friction that is inconceivable unless one has experienced war'.[71] Although it was in *On War* that Clausewitz best articulated the idea of friction and chance, it was not a new idea. In his historical account of Napoleon's Russian campaign he writes:

> The instrument of war resembles a machine with prodigious friction, which cannot, as in ordinary mechanics, be adjusted at pleasure, but is ever in contact with a host of chances. War is, moreover, a movement through a dense medium.[72]

---

70   Gat, *Military Thought*, 168; See also: Clausewitz, 'On the Life and Character of Scharnhorst', 100. Although references to theoretical influences are extremely rare in *On War*, Clausewitz does refer to Scharnorst's use of history, noting that 'Sharnhorst, whose manual is the best that has ever been written about actual war, considers historical examples to be of prime importance'. Clausewitz, *On War*, 170.

71   Clausewitz, *On War*, 138.

72   Clausewitz, *The Campaign of 1812 in Russia*, 185.

Theoretical Influences

Despite the length and depth of Clausewitz's tome, there is surprisingly little reference to other writers, and Clausewitzian scholars continue to speculate about the influences on Clausewitz's work. We know with certainty that Clausewitz used historical evidence as the basis for his theory. He had witnessed firsthand the transformational changes in the character of war which had brought Prussia to its knees. Yet, how did he utilize the theoretical works of other scholars? Although there is a general consensus that Clausewitz used a range of philosophical methods in his own approach, he provides little indication of which thinkers influenced his ideas. In one of his rare references to another writer, he reveals that:

> My original intention was to set down my conclusions on the principle elements of this topic in short, precise, compact statements, without concern for system or informal connection. The manner in which Montesquieu dealt with his subject was vaguely in my mind [73]

Did Montesquieu provide the template for Clausewitz's reasoning in *On War*? According to Ulrike Kleemeier, Clausewitz's appreciation of the role of emotion connects him closely with philosophers like Montesquieu, Spinoza, Hobbes and Hume, 'not with Hegel and still less with Kant'.[74] There is a certain similarity between Montesquieu's use of historical and empirical evidence in the pursuit of universal truths and Clausewitz's own method of inquiry. Does this link him to Montesquieu? The easy answer is yes; in fact, there is plenty of evidence which links him to a variety of thinkers. What is harder to gauge is the extent of their influence on Clausewitz's theory. As Azar Gat reminds us, Clausewitz pointed out that Montesquieu was 'only vaguely in my mind'.[75] Although Clausewitz was clearly aware of Montesquieu, he had a wide-ranging interest in politics, philosophy and history and appears to have drawn influences from a diverse range of work. From existing documents we know that he had correspondence with the German philosopher J.G. Fichte, in which Clausewitz comments in depth on the political lessons from Machiavelli's writings. We can also extrapolate with some certainty that he gleaned his original insights regarding the political nature of war from Machiavelli. The idea was not new and had been postulated by various thinkers before Clausewitz – Machiavelli and Guibert not least.[76]

---

73 Clausewitz, 'Note of 1818', 71.

74 Ulrike Kleemeier, 'Moral Forces in War', in Strachan and Herberg-Rothe, *Clausewitz in the Twenty-first Century*, 109.

75 Gat, *A History of Military Thought*, 194.

76 Clausewitz, 'Letter to Fichte (1809)', in Clausewitz, *Historical and Political Writings*, 279–84. The timing of the letter (1809) corresponds with Clausewitz's more strident views before the French disasters in Russia and he celebrates Fichte's interpretation and the Prussian need to exist in a self-help system of states struggling to regain equilibrium.

As his theoretical evolution continued, so he encountered a range of philosophical works which are thought to have infused *On War* with greater depth, and which complicate the articulation of his ideas. Indeed, while he is linked with thinkers such as Machiavelli and Montesquieu, he is most commonly associated with Kant, and to a lesser extent Hegel.[77] For some, the influence of Hegelian dialectics with its emphasis on thesis, anti-thesis, and synthesis, reflects the Trinity in particular.[78] Gat comments that part of the lure of Hegel may have been that 'one of the chief lessons of this philosophy was that all the contrasts and contradictions of reality were actually but different aspects of a single unity'.[79] Hegelian dialectics may have offered a way out of the impasse Clausewitz found himself in when he began his revision in 1827, bringing unity to his ideas. The divergent strands of his arguments did not mean that war had no nature; simply that it is stochastic and complex in its composition – war's separate components could be brought back into a wider whole – political interaction.

The connections with Kant are also hard to ignore. According to Clausewitz's principle biographer, Peter Paret, he never studied Kantian or Hegelian philosophy and gained knowledge of them third hand through public lectures and pamphlets.[80] He certainly had access to their theoretical works at the war college, and as Echeverria has shown, his understanding of Kantian logic appears to have derived from the lectures on the '*Outline of General Logic*' taught by Johann Gottfried Kiesewetter.[81] In addition, we know from early letters that Clausewitz was interested in study in order to find truths as well as simply win wars, and his access to German philosophers, principally Kant, invested him with his ability in 'speculative reasoning'.[82] Although a son of the Enlightenment, Clausewitz rejected purely positivist explanations that war is governed by set laws, and instead utilized counter-enlightenment ideas, such as Romantic idealism, especially the importance 'placed on the psychological qualities of the individual', which is axiomatic to his ideas on military genius and the role of emotion and intuition.[83] Although the Enlightenment's rationalism was transforming the way in which society approached the intellectual challenges of the period, as Clausewitz remarks in *On War*, a significant problem with this kind of approach was that rationalist theorists 'directed their inquiry exclusively toward physical quantities,

---

77   Clausewitz's correspondence with Fichte, primarily concerning Machiavelli and especially the methods of educational theorist Johann Heinrich Pestalozz provides insight regarding his own ideas on building and disseminating knowledge. See: Paret, *Clausewitz and the State*, 169–208.

78   Christopher Bassford, *Clausewitz and His Works* (2002). www.clausewitz.com/CWZSUMM/CWORKHOL.htm (accessed 14 September 2010).

79   Gat, *A History of Military Thought*, 234.

80   Peter Paret, 'Clausewitz', in Peter Paret (ed.), *Makers of Modern Strategy* (Princeton: Princeton University Press, 1986), 194.

81   Echevarria, *Clausewitz and Contemporary War*, 22.

82   Paret, *Clausewitz and the State*, 149.

83   Paret, *Clausewitz and the State*, 149.

whereas all military action is intertwined with psychological forces and effects'.[84] As Clausewitz explains, 'Military activity is never directed against material forces alone; it is always aimed simultaneously at the moral forces which give it life, and the two cannot be separated'.[85]

Clausewitz's 'continuous dialogue' is an example of his search for absolute truth, as exemplified by Kantian logic. This penetrates to the very heart of Clausewitz's theory. It becomes all the more apparent when we return to the aim of his study, especially the tension between purpose and means. If 'war is an act of force to compel our opponent to fulfil our will', what is the course of action which best achieves this aim? This is the basic problem that Clausewitz is faced with and it is from this starting point that the rest of his ideas originate. His writing is layered with Kantian reasoning, particularly the tension between 'logical' and 'material' truth.[86] We can see this reasoning as the very idea that caused Clausewitz to revise his whole treatise: that war can be of two kinds. While he rejected many of the pretensions of the rationalist thinking of his era, neither did he completely embrace purely non-rationalist arguments, articulated by those such as Berenhorst in his *Reflections on the Art of War*.[87] Clausewitz often chose to draw his methodological approach from competing dialectical arguments; drawing on, but balancing between competing visions.[88] It was this interaction between theory and experience that enabled him to convey his argument that defence, rather than attack, is the stronger form of war.[89] This insight was gleaned through experience relating directly to the wars of his period. The power of the offensive could be overplayed, as it was by Napoleon in Russia and in Spain. Although attack has a positive connotation, Clausewitz notes: 'The attacker is purchasing advantages that may become valuable at the peace table, but he must pay for them on the spot with his fighting forces.' He continues:

> There are strategic attacks that have led directly to peace, but these are the minority. Most of them only lead up to the point where the remaining strength is just enough to maintain a defence and wait for peace. Beyond that point the scale turns and the reaction follows with a force that is usually much stronger than that of the original attack. This is what we mean by the culminating point of attack.[90]

---

84   Clausewitz, *On War*, 156.

85   Clausewitz, *On War*, 157.

86   Echevarria offers an excellent summary of Clausewitz's use of logical and material truth – as taught by Kiesewetter in his lectures on the 'Outline of General Logic'. Echevarria, *Clausewitz and Contemporary War*, 21–54.

87   See, Paret, *Clausewitz and the State*, 149.

88   Gat, *Military Thought*, 167.

89   For an excellent exposition of this Clausewitzian theme, see: Jon Sumida, 'On Defence as the Stronger Form of War', in Strachan and Herberg-Rothe, *Clausewitz in the Twenty-first Century*, 162–81.

90   Clausewitz, *On War*, 639.

In Chapter Twenty-five of Book Six on Defence, he highlights his reasoning; a direct comparison with Napoleon's Russian campaign is hard to ignore:

> We regard a voluntary withdrawal to the interior of the country as a special form of indirect resistance – a form that destroys the enemy not so much by the sword as by his own exertions. Either no major battle is planned, or else it will be assumed to take place so late that the enemy's strength has already been sapped considerably.[91]

Of course, Clausewitz is particularly famous for theorizing about the escalatory nature of war. He hypothesized that, 'If one side uses force without compunction, undeterred by the bloodshed it involves, while the other side refrains, the first will gain the upper hand. That side will force the other to follow suit; each will drive its opponent towards extremes'.[92] He continues: 'To introduce the principle of moderation into the theory of war itself would always lead to logical absurdity … This is the *first case of interaction and the first 'extreme'* we meet with.'[93] In this reciprocal contest Clausewitz continues to proceed logically up an escalatory ladder caused by reciprocal interaction, highlighting the *second and third cases of interaction and the extreme*. In the *second interaction leading to the second extreme*, he concludes that 'if you are to force the enemy, by making war on him, to do your bidding, you must either make him literally defenceless or at least put him in a position that makes this danger probable'. *The third interaction leading to the third extreme* relates to the force used to overcome one's opponent: 'if you want to overcome your enemy you must match your effort against his power of resistance' – you must employ the maximum use of force. In other words, the extremes drive the tendency to reciprocal violence; 'and there is no logical limit to the application of that force'.[94]

This is a primary example of Clausewitz's abstract theorizing, a legacy of his third-hand Kantian tutelage. If one is to compel the enemy to one's will, then it makes sense to use all the power at one's disposal. Logically there would be no restriction on violence due to the natural escalatory proclivity that war manifests. However, when Clausewitz tests this logical truth against material truth – historical examples – he finds that war in its ideal abstract form is unachievable. It exists in theory only. Friction, chance, purpose and politics all conspire to limit war's violent tendencies. As Clausewitz makes clear in his 'authors' preface', written between 1816 and 1818,

---

91   Ibid., 566.
92   Ibid., 84. Andreas Herberg-Rothe offers an excellent illustration of the 'interactions to the extreme' and how they are limited by 'real' world problems. See Herberg-Rothe, *Clausewitz's Puzzle*, 39–52.
93   Clausewitz, *On War*, 84–5.
94   Ibid., 85.

Just as some plants bear fruit only if they don't shoot up too high, so in the practical arts the leaves and flowers of theory must be pruned and the plant kept close to its proper soil – experience.[95]

In the wake of Clausewitz's decision to revise his treatise, he re-evaluates and re-emphasizes politics as an overarching synthesis. This provided his theory with clarity without undermining the divergent forces that Clausewitz argued made war such a unique human activity. As Echevarria has pointed out in relation to the Trinity, Clausewitz could have easily arrived at the position that war has no universal nature – the concept reminds readers not to fix an arbitrary relationship between the three tendencies.[96] War seems to exhibit divergent tendencies. It is the political umbrella which provides the unifying idea capable of bringing his divergent ideas together in a meaningful way. It is the overarching political synthesis that provides unity to his Trinitarian concept, as it does his treatise as a whole.[97] This is entirely in keeping with Clausewitz's early thoughts and his own intuitive desire to systematize his findings. In the note dated 1816–18, he states that he intends to leave the reader with 'small nuggets ... That is how the chapters of this book took shape, only tentatively linked on the surface'.[98]

As Clausewitz himself highlighted, he began his study by examining separate phenomena only to find that individually these independent 'kernels' were insufficient if he was to develop a coherent understanding of the nature and conduct of war. Furthermore, as Paret has argued, Clausewitz did not build his theory by using the methodology of any one approach. His approach resembles a pastiche of methods and ideas – he took and used what seemed useful in his search for greater intellectual and practical clarity. He was a soldier, not a philosopher.[99] By balancing philosophical and historical methods against his own experience of warfare he came not only to conceive of two types of war, but also that the different strands of thought which he had been studying, when juxtaposed against experience, brought him to the conclusion that war was comprised of a 'Wondrous Trinity'. From as early as the 1816–18 note, he tells us that 'Analysis and observation, theory and experience must never disdain or exclude each other; on the contrary, they support each other'.[100] Experience and empirical evidence perform a key function in his methodology, testing the validity of his conceptualizations against the hard facts of the real world. He dedicates the whole of Chapter Six of Book Two, 'On Historical

---

95   Clausewitz, 'Author's Preface – To an Unfinished Manuscript on the Theory of War, written between 1816 and 1818', *On War*, 69.

96   Echevarria, *Clausewitz and Contemporary War*, 62–83.

97   Clausewitz, 'Note of 1827', 77.

98   Ibid., 70.

99   Paret, *Clausewitz and the State*, 150.

100   Clausewitz, 'Note 1816–1818', 69.

Examples', to the use of military history as a source of empirical knowledge: the evidence he needs to test his ideas.[101]

Based on empirical evidence, *On War* elucidates and debates the influence of factors outside the purview of the writings of his contemporaries. Clausewitzian theory does not fix immutable rules, but reflects on the infinite number of imponderables of real life, and on how these influence war. It was this method, however messy, that resulted in the 'Wondrous Trinity' – it reflects his whole approach to theory. It is in his Trinitarian formula of interplay between 'hostility, chance, and reason' that Clausewitz draws the competing strands of thought into a workable, explanatory model. The individual components of the concept are first explored separately, and finding that the impact of each tendency cannot be understood fully without assessing their connections with the others, they are unified in his tri-part concept. Clausewitz tells us that his work 'consists of an attempt to investigate the essence of the phenomenon of war and to indicate the links between this phenomenon and the nature of its component parts'.

---

101   Clausewitz, *On War*, 199–204.

Chapter 2
Clausewitz and the Tale of Two Trinities

Clausewitzian Studies and the 'Trinity'

Although the Trinity has become fashionable, and recent scholarship has undoubtedly resulted in major advances in the way the concept is perceived, it has not ventured either to investigate the contradiction between war as a Trinity, and war as a rational instrument of policy, or test the concept empirically. This book fills that gap, examining the tension between Clausewitz's ostensibly competing ideas. Despite the richness of contemporary Clausewitzian scholarship, those focusing on the Trinity claim that its components are essentially equal, in the process undermining Clausewitz's political argument and the very idea that has imbued his writings with such enduring significance.

The current chapter examines this tension more fully. Its purpose is to tease out the strengths and weaknesses of the concept and more fully understand its position within Clausewitzian thought. Although the seeds of this discussion have been laid in the introductory chapter, the present explores the topic more fully. It initially charts our evolving understanding of the concept within the strategic studies community, and not least the reasons for its misinterpretation and successive dismissal by advocates of the new war thesis. The chapter subsequently examines the 'core' Trinity in detail, and explores the insights and complexities of the concept. In the following section it reassesses the role of the Trinity and its relationship to Clausewitz's claim that war is a continuation of politics.

Clausewitz: And the Tale of Two Trinities

While the Howard/Paret translation remains the most accessible text available, its interpretation may be inadvertently responsible for the Trinity, when re-discovered, being linked not to the core Trinity, but to the people, army, and government. As Jan Willem Honig has highlighted, earlier translations had painted a picture of Clausewitz as a militarist, the Howard/Paret version on the other hand presents Clausewitz as a liberal scholar working in the search of objective knowledge, uninfluenced by the social conventions and prejudices of his own period. The *On War* of Howard and Paret promotes Clausewitz's ideas as essentially rational, able to inform the era's policy makers with valuable insights. Honig remarks:

> Their Clausewitz is a man therefore whom, in critical respects, they regard as a
> modern contemporary; a man who speaks directly to us and who, moreover, has
> pertinent things to say about the major challenge of the time when the translation
> was made, the Cold War.[1]

The core Trinity of hostility, chance, and policy/reason has played second fiddle
to the more commonly known people, army, and government model. Although a
discussion of the Trinity and the new war challenge has already taken place in the
introductory chapter, it is important to re-examine why the Trinity is so distorted in
the existing literature. The following section briefly examines this misconception,
before then exploring how the study of the Trinity has been shaped by the recent
new war debate.

Although the Trinity is currently in vogue, it has not always attracted so much
attention. As one Clausewitzian scholar remarked, had he got round to revising
the whole treatise, it may 'have opened Clausewitz's eyes to certain weaknesses
– including the near-banality of the conclusion – of his revised opening
chapter'.[2] While this may appear a bit harsh, the point is that debate regarding
the Trinitarian aspect of Clausewitz's work has been largely overlooked and the
present interest in the concept is relatively new. In fact the re-discovery of the
Trinity as a core element in Clausewitzian thought is largely attributable to US
introspection following the Vietnam War. The second Trinity – the people, army,
government model – as propounded by Van Creveld, Kaldor, and Keegan, came
to prominence following the publication of the late Colonel Harry Summers'
study, *On Strategy: A Critical Analysis of the Vietnam War*.[3] At a time when
the United States was struggling to understand its strategic defeat at the hands
of the Communist Vietnamese forces, Summers' analysis presented an account
that clearly defined where the US had gone wrong. It seemed to identify the
cause of America's strategic reverse in Vietnam. In the context of the era, the
demoralizing anti-war movement, political infighting and lack of strategic
direction seemed an obvious reason for US military failure. Summers' findings
reflected this, presenting clear guidelines for future operations. The lesson
to be drawn from this unfathomable defeat, at the hands of a technologically
inferior enemy, was, according to Summers, to generate a better awareness of
the principles propounded by Clausewitz. For Summers, using the Clausewitzian
Trinity of people, army, and government, was paramount if the US military was
to refocus its efforts on future threats – most notably the potential clash with the
Soviet Union in Europe. In Summers' words, 'The task of the military theorist,

---

1 Jan Willem Honig, 'Clausewitz's On War: Problems with Text and Translation', in
Clausewitz in the Twenty-First Century, 60.

2 W.B. Gallie, *Philosophers of Peace and War: Kant, Clausewitz, Marx Engels and
Tolstoy* (Cambridge: Cambridge University Press, 1978), 48.

3 Col. Harry, G. Summers Jr, *On Strategy: A Critical Analysis of the Vietnam War*
(Novato, CA: Presidio Press, 1982).

Clausewitz said, is to develop a theory that maintains a balance among what he calls a trinity of war – the people, the government, and the army'.[4] This analysis mirrored the wider mood in the US military at the time.

In the same period as Summers was emphasizing the importance of the Trinity – albeit the wrong one – US military interest in Clausewitz's ideas was generated even further by the intellectual contributions of Colin Powell and Caspar Weinberger. The result of this interest culminated in the 'Weinberger Doctrine' – which prohibited the US military from military action unless exacting criteria specifying how the US national interest was served by the use of force could be met.[5] Weinberger proclaimed:

> Policies formed without a clear understanding of what we hope to achieve would also earn us the scorn of our troops, who would have an understandable opposition to being used – in every sense of the word – casually without intent to support them fully. Ultimately this course would reduce their morale and effectiveness for engagements we must win.[6]

In other words, there must be achievable political aims. Of course, this resonates of the Clausewitzian claim that 'war is the continuation of politics'. The architects of the doctrine sought to cement the US military back into a rational setting after the debacle of Vietnam. As Strachan observes, 'Powell and Weinberger were attracted to Clausewitz precisely because he seemed to be so clear about the relationship between war and policy'.[7] Vietnam had shown the dangers of incremental escalation, and the US would not be sucked into a war without good reason. Military adventures of this type had the potential to get bogged down and lose public support. By binding defence strategy to clear principles for action, the US defence community sought to immure future policy decisions into a crystallized Clausewitzian formula that would produce predictability of outcome; an understandable though ultimately unattainable wish. As one analyst put it at the time, 'Given that national survival may hinge on the ability to deter or win wars,

---

4   Summers, *On Strategy: A Critical Analysis of the Vietnam War*, 26–7. This argument is reproduced in Summers' study of the Gulf War: *On Strategy II: A Critical Analysis of the Gulf War* (New York: Dell, 1992), 11.

5   Kinross, 'The Influence of Clausewitz upon American Strategic Thought', 183–6. The six key tenets of the doctrine were: (i) The vital interests of the United States or its allies must be at stake; (ii) Once the decision to fight is taken the effort should be wholehearted so as to reflect the intention of winning; (iii) political and military objectives must be clearly defined; (iv) political and military objectives must be continuously reassessed to maintain balance between ends and means; (v) before troops are sent abroad, there must be some reasonable assurance of Congressional and public support; (vi) the commitment of US forces to war should be undertaken only as a last resort.

6   Casper W. Weinberger, *Fighting For Peace: Seven Critical Years in the Pentagon* (New York: Warner Books, 1990), 437.

7   Hew Strachan, *Clausewitz's On War* (Atlantic Monthly Press, 2006b), 3.

the value of discovering enduring principles of strategic success is obvious'.[8] Clausewitz's Trinity appeared to fill that gap. As Weinberger put it, never again should the imperative of public support be ignored.'[9] He continues:

> Any U.S. government that attempts to fight where our vital interests are not at stake, when we have no good reason to suppose there will be continuing public support, committing military forces merely as a regular and customary adjunct to our diplomatic efforts, invites the sort of domestic turmoil we experienced during the Vietnam War. Such a government has no grounds for expecting any less disastrous result.[10]

Like Summers, Powell – the future Chairman of the Joint Chiefs of Staff – and Caspar Weinberger – the then Secretary of Defense – had misinterpreted Clausewitz's message. For Powell in particular there was an emphasis on maximizing US 'conventional' firepower so it was better able to target the 'Centre of Gravity' of any would-be opponents. It was a decision which seemed prescient when the US and its allies unleashed the firepower and tempo which easily broke the resistance of the Iraqi army in 1991. We can see the logical culmination of the so-called Powell doctrine again during operation 'Shock and Awe' which formed the opening salvos of the US invasion of Iraq in 2003.[11] For Powell, and many others within the US military, 'conventional' war-fighting dominance was a piece of the jigsaw, thus complementing the other supposed elements of the 'Trinity' – maximizing the cohesion of the triad of people, army, and government was the key to victory. With such patronage, it is not surprising that Clausewitz's ideas, despite being distorted, have been enshrined in US army doctrine; as they are in British equivalents. For Summers, Powell, and Weinberger, and for a great many others within the US military – past and present – the Trinity of people, military, and government represented the formula for strategic success. As Powell put it,

> Clausewitz's greatest lesson for my profession was that the soldier, for all his patriotism, valour, and skill, forms just one leg in a triad. Without all three legs

---

8   Stephen M. Walt, 'The Search for a Science of Strategy: A Review Essay', *International Security*, 12 (Summer, 1987), 141.

9   Casper W. Weinberger, 'U.S. Defense Strategy', *Foreign Affairs* (Spring, 1986), 685.

10   Weinberger, 'U.S. Defense Strategy', 689.

11   The patronage of Weinberger and Powell ensured Clausewitz's importance to the US military establishment. Increasingly, however, Clausewitz's intended message has been diluted and even replace by the entrenched assumption that Clausewitzian theory propounds an all-out military effort using the best and newest technologies. Overwhelming power has been consistently but wrongly identified as the key to strategic success.

engaged, the military, the government, and the people, the enterprise cannot stand.[12]

Although the people-army-government model has undoubtedly deviated from Clausewitz's meaning, as noted already, it was Clausewitz himself who married the forces inherent in his core Trinity: passion, hatred and enmity with the people; the play of chance and probability with the army; and war's subordination to reason with the government. However, this linkage was not intended as an explicative exemplar which set fixed markers; as illustrated previously, such an approach contradicts the *raison d'être* of his entire work.[13]

For Clausewitz, reducing war to a few scientific principles was a futile exercise, not least because it didn't assess the role of psychology or account for war's reciprocal nature – his contemporaries failed to consider that one's opponent had a free will which would inevitably clash with one's own – a core insight of the Trinitarian nature of war In contrast, *On War* is concerned with the search for verifiable truths about war, not only in Clausewitz's own period, but universally, throughout the totality of history: past, present, and future. He was looking for the universal elements of war and as a result rejects theories which focus on parts rather than the whole. He notes in Book Two, Chapter Two:

> Formerly, the terms 'art of war' or 'science of war' were used to designate only the total body of knowledge and skill that was concerned with material factors. The design, production, and the use of weapons, the construction of fortifications and entrenchments, the internal organization of the army, and the mechanism of its movements constituted the substance of this knowledge and skills ... It was about as relevant to combat as the craft of the sword-smith to the art of fencing. It did not yet include the use of force under conditions of danger, subject to constant interaction with an adversary, nor the efforts of spirit and courage to achieve a desired end.[14]

---

12   Colin Powell, with Joseph Persico, *My American Journey* (New York: Random House, 1995), 207–8.

13   Actually, the arguments of Van Creveld, Keegan or Kaldor, which suggest the people, army, government model is no longer relevant are highly debatable. For instance, Isabelle Duyvesteyn's, *Clausewitz and African War: Politics and Strategy in Liberia and Somalia* has demonstrated the continued validity of the people, army, and government paradigm in ostensibly new wars. In all wars, there will be a people, army, and government of sorts. See: Isabelle Duyvesteyn, *Clausewitz and African War: Politics and Strategy in Liberia and Somalia* (Abingdon: Frank Case, 2005). This view is also shared by Christopher Bassford and Edward J. Villacres, 'Reclaiming The Clausewitzian Trinity', *Parameters* (Autumn, 1995), 9–20.

14   Clausewitz, *On War*, 153.

The prescriptive formula with which Summers, Powell, and Weinberger accredited Clausewitz would completely negate On War's *raison d'être*. While the model propounded by writers such as Summers and Powell typifies the notion of strategic rationality, the core Trinity – hostility, chance, and policy – explicates a phenomenon which is interactive, reciprocal and beguiling at best. That modern war displays irrational proclivities is not something terribly unusual, nor can it be confined to war in one particular period. Clausewitz understood only too well that in all its complexity, war can reveal irrational tendencies. As Michael Handel remarked of Clausewitz:

> It has often been argued that Clausewitz emphasizes the need to view war as a rational instrument, as a means for the leaders to promote and protect their state's vital interests. From this accurate interpretation, however, some readers have erroneously inferred that Clausewitz also considers it possible for *war itself* to be waged as a rational activity. In fact, Clausewitz, repeatedly reminds us that this is not so, for he knows that war in all of its dimensions is permeated by non-rational influences, or what he calls 'moral factors' (*moralische Grossen*), 'spiritual forces' (*geistigen Kraften*), or 'spiritual factors' (*geistigen Grossen*), which 'cannot be classified or counted'.[15]

Of course, the idea that war is a result of political action comes directly from the pages of his opus. As Clausewitz himself put it, 'war is not merely an act of policy but a true political instrument, a continuation of political intercourse, carried on with other means'.[16] For Clausewitz, the symbiotic relationship between war and politics stems from the very essence of what conflict is – it plays a vital and functional role. As he is at pains to remind readers, war is 'an act of violence intended to compel our opponents to fulfil our will'.[17] From this starting position he understood war as a continuation of policy. He was cognizant that if war is intended to compel an enemy to accept your will, it should be remembered that 'the political object is the goal, war is the means of reaching it, and means can never be considered in isolation from their purpose'.[18] Clausewitz may have famously conceptualized an 'ideal' type of war with no limits on the levels of violence; however, he was acutely aware that this theoretical 'ideal' existed as an abstract, detached from the 'real' world. Though violence drives violence, real war is restricted by political aims – and the physical ability to coerce one's opponent. At some point war must be curtailed by policy. If it is not, then it has become something other than war.[19] Colin Gray reiterates this point:

---

15   Michael I. Handel, *Masters of War*, third edition (London: Frank Cass, 2001), 81.
16   Clausewitz, *On War*, 99.
17   Clausewitz, *On War*, 81.
18   Clausewitz, *On War*, 99.
19   Clausewitz, *On War*, 83–4.

Some confused theorists would have us believe that war can change its nature. Let us stamp on such nonsense immediately. War is organized violence threatened or waged for political purposes. That is its nature. If the behaviour under scrutiny is other than that just defined it is not war.[20]

In terms of his Trinity, Clausewitz intended his famous concept to act as a model with which to comprehend the complexity of war once battle has commenced. When one understands that war is shaped by the interplay of complex forces – hostility (passion and hatred), chance and uncertainty, and policy (rationality) – it is clear that strategic calculations must be constantly re-correlated to account for 'ends and means'. In short, what is the political aim of the conflict, and how do we reach a suitable outcome? As war is a reciprocal activity, its fluid and unpredictable nature ensures not just that prescription is difficult, but that prediction about the outcome of a particular war founded upon anything other than political dexterity will sooner or later end disastrously. In what has been described by one Clausewitzian scholar as the most important input into the debate about Clausewitz, Alan Beyerchen brilliantly develops Clausewitz's understanding of war – as expressed by his Trinity – as a reflection of non-linearity.[21] Beyerchen states:

I am not arguing that reference to a few of today's 'non-linear science' concepts would help us clarify confusion in Clausewitz's thinking. My suggestion is more radical: in a profoundly unconfused way, he understands that seeking exact analytical solutions does not fit the non-linear reality of the problems posed, and hence that our ability to predict the course and outcome of any given conflict is severely limited.[22]

Comprehension of such complexity in turn focuses attention towards finding a suitable strategy, one which needs constant reflection and adaptation. Of course, the purpose of *On War* is to provide soldiers with the insights which might enable them to produce good strategy. The Trinity is the starting position from which this can be achieved. As Christopher Bassford has put it, the Trinity 'is the concept that ties all of Clausewitz's many ideas together and binds them into a meaningful whole'.[23]

---

20   Colin S. Gray, *Another Bloody Century* (London: Weidenfeld & Nicolson, 2005), 30.

21   Alan D. Beyerchen, 'Clausewitz, Nonlinearity and the Unpredictability of War', *International Security*, 17(3) (1992), 59–90.

22   Ibid, 61.

23   Christopher Bassford, 'Tip-Toe Through the Trinity, or The strange persistence of Trinitarian warfare, www.Clausewitz.com/CWZHOME/TRINITY/TRINITY8.htm, (2007), 4 (accessed 2 January 2009). An earlier version is published as: Christopher Bassford, 'The Primacy of Policy and the "Trinity" in Clausewitz's Mature Thought', in Hew Strachan and Andreas Herberg-Rothe (eds), *Clausewitz in the Twenty-first Century* (Oxford: Oxford University Press), 74–90.

While the new war challenge questioned the efficacy of the Trinitarian concept, it did so because of a misperception that Clausewitz premised his treatise on state power – considered to be 'rational' because of Clausewitz's own instruction that war should be considered a political instrument. The new war rejection of Clausewitzian theory therefore rests on two misconceptions. Firstly, Clausewitz was also cognizant of other types of war than those waged with standing state armies. Secondly, *On War* exemplifies the multifarious and capricious; it shies away from didacticism and considers prescriptive formulae a misnomer. Though much of the debate surrounding modern war stems from debate about the role of the Trinity, few of the new war writers can adequately demonstrate an appreciation of the core concept. Most simply aver that Clausewitzian theory is premised on a position of state primacy.[24]

Moving on from the new war challenge, just as there has been a proliferation of new war adherents there is also a growing corpus of literature supporting the position propounded by Clausewitz. Scholars such as Colin Gray, Christopher Bassford, Alan Beyerchen, Antulio J. Echevarria, Hew Strachan and Andreas Herberg-Rothe, among others, have revitalized Clausewitzian scholarship in response to the new war polemic.[25] Building on a paper delivered in 2005 to an Oxford University conference on 'Clausewitz in the 21st Century', Christopher Bassford's working draft, 'Tip-Toe Through The Trinity', demonstrates the relevance of the Trinitarian concept by returning to the original text of Clausewitz's *On War*.[26] Our comprehension of Clausewitz's true meaning of the Trinity has been aided even further by the meticulous analysis given to the subject by Echevarria's *Clausewitz and Contemporary War*, and Andreas Herberg-Rothe's excellent *Clausewitz's Puzzle*, both of which rightly place the concept at the heart of Clausewitz's opus. By reflecting on Clausewitz's original arguments, writers such as Bassford, Echevarria, and Herberg-Rothe have been able to better represent *On War*'s message. As a result they have successfully demonstrated that the core of the Trinitarian concept is in fact comprised of (i) hatred passion and enmity, (ii) the play of chance and probability and (iii) war's subordination to rational policy, rather than the people, army, government model favoured in the wider new war and traditional strategic studies literature.[27]

---

24   Colin M. Fleming, 'New or Old Wars? Debating a Clausewitzian Future', *The Journal of Strategic Studies*, 32(2) (April, 2009).

25   Important works include: Beyerchen, *Clausewitz, Nonlinearity and the Unpredictability of War*; Echevarria, *Clausewitz and Contemporary War*, Hew Strachan, *Clausewitz's On War*; and Andreas Herberg-Rothe, *Clausewitz's Puzzle* (Oxford: Oxford University Press, 2007).

26   The proceedings of the conference are now available as an edited volume. See: Hew Strachan and Andreas Herberg-Rothe (eds), *Clausewitz in the Twenty-first Century.*

27   Like Bassford, Handel and Echevarria support the continued use of the Trinitarian concept. However, even supporters of Clausewitz disagree on its exact use. While Handel argues that the Trinity should be 'squared' to account for the role of technology, Echevarria, like Gray and Bassford, claim that technology does not undermine the original concept.

There is thus reason to believe that this 'real' Trinity is universally applicable, and can therefore be used to analyse war in the modern world. Though the second model – people, army, government – is widely thought to reflect a state-centric understanding of war, and is thus open to criticism in a world where war can be fought increasingly by a range of actors, the core Trinity championed by Bassford, Echevarria, and Herberg-Rothe, should account for war between any variety of actors. Crucially, by building on the important work of Alan Beyerchen, these scholars have highlighted the non-linear focus of Clausewitzian theory. They have revivified Clausewitz's ideas just when international politics regularly displays its non-linear characteristics, and when a proliferation of non-state actors are able to wage war. According to these 'neo-Clausewitzians', the Trinity is not a staid expression of state war from the Napoleonic period; rather, it expresses the very complexity of war itself.

The True 'Trinity': Hostility, Chance, and Policy

> There is something deeply perplexing about the work of Carl von Clausewitz. In particular, his unfinished magnum opus *On War* seems to offer a theory of war, at the same time that it perversely denies many of the fundamental preconditions of theory as such – simplification, generalisation and prediction, among others.[28]

Although *On War* elucidates complexity rather than simplification, the refusal to proscribe immutable truths to the subject is a central reason why Clausewitz stood, and stands, apart from his contemporaries or would-be modern competitors. His 'Wondrous Trinity' forms his explication of war's complex form. It was through this concept that Clausewitz brought the different strands of his thought together and where he articulates war's inherently problematical nature. As such, Beyerchen's statement above can hardly be refuted. In fact, Beyerchen is not casting doubt on Clausewitz's ideas. Instead, he links them brilliantly to a focus on non-linear science and its application to understanding *On War.*

Linked to chaos theory, non-linearity begins to elucidate the sheer complexity of war. Using chaos and complexity theory as a reference point, it suggests that even the slightest of alterations and changes of focus, either intentionally or otherwise, produces a cause-and-effect relationship which sends out 'ripples'. Starting off as unimportant, these same 'ripples' are influenced by yet more alterations, causing the

---

A very good account of this debate can be found in: Antulio J. Echevarria II, 'War and Politics: The Revolution in Military Affairs and the Continued Relevance of Clausewitz', *Joint Forces Quarterly* (Winter, 1995), 76–82.

28   Alan D. Beyerchen, 'Clausewitz, 'Nonlinearity and the Unpredictability of War', *International Security*, 17(3) (1992), 59.

small and often insignificant problems to expand exponentially.[29] This cause-and-effect relationship is very different to that commonly associated with Clausewitzian thought today. As Clausewitz was cognizant, causes may have effects but these will probably be other than expected. In war, any action and counter-reaction moves the original strategy away from its intended goal. In other words, we should understand war as a constantly moving, evolving phenomenon, different in every case, yet universal in its core elements. Unless the strategist has a clear grasp of this fact, and is able to undertake the constant re-correlations between ends and means, disaster will be a constant companion. This is the message evident throughout *On War*; it is conveyed as a general theory in the Trinity.

Clausewitz's Trinity displays complexity rather than simplification; it is an idea which runs through the entirety of his opus and is rooted in his appreciation that war is a multi-lateral, reciprocal and vitalistic activity. From the very first paragraph Clausewitz emphasizes the importance of considering the 'whole'. From this starting point he introduces reciprocity: 'War is nothing but a duel on a larger scale ... but a picture of it as a whole can be formed by imagining a pair of wrestlers. Each tries through physical force to compel the other side to do his will.'[30] War on paper may appear easy, but war in the real world, impacted by chance, friction and not least by passion and hostility, is decidedly more fickle. He explains:

> The essential difference is that war is not an exercise of the will directed at inanimate matter, as is the case with the mechanical arts, or at matter which is animate but passive and yielding, as is the case with the human mind and emotion in the fine arts. In war, the will is directed at an animate object that reacts.[31]

When the multi-layered forces in war are conflated by reciprocity, the result is uncertainty; war is a fluid, unstable, and non-linear activity. Furthermore, better comprehension of that complexity and instability is the starting point if strategy is to use war instrumentally, as 'a continuation of policy'. This is not an easy task. As Clausewitz remarked, 'Bonaparte was quite right when he said that Newton himself would quail before the algebraic problems it could pose.'[32]

As was highlighted in the previous chapter, Clausewitz's methodology comprised a mixture of philosophical ideas, tested against empirical evidence.

29　Important works on chaos and complexity theory include: James Gleick, *Chaos: Making a New Science* (New York: Viking, 1987); Rodger Lewin, *Complexity: Life at the Edge of Chaos* (London: Pheonix; new ed, 2001); and Mitchell, M. Waldrop, *Complexity: The Emerging Science at the Edge of Order and Chaos* (New York: Simon and Schuster, 1992).

30　Clausewitz, *On War*, 83.

31　Clausewitz, *On War*, 173–4.

32　Clausewitz, *On War*, 708.

For much of his examination the different strands of his study remain separate. Nevertheless, in his 'attempt to investigate the essence of the phenomena of war and to indicate the links between these phenomena and the nature of their component parts,' he began to systematize his work.[33] This inclination to systematize and his desire to understand war's vitalism brought him to unify the divergent threads of his investigation under a political theme – war as a continuation of politics. The three parts of the Trinity could only be understood as a whole if they were first united by a common idea. That idea was politics.

The three component parts of the Trinity are individually important, but it is only by grasping the degree to which they interact that one will be able to appreciate war's complexity. The core Trinity – hostility, chance, and reason – which is presented in detail below stands in clear contradistinction to the pre-fixed and staid markers of the people, army, and government model. Marking the conclusion to Chapter One, Book One, Clausewitz introduces his Trinity. He informs us that:

> War is more than a true chameleon that slightly adapts its characteristics to the given case. As a total phenomenon its dominant tendencies always make war a paradoxical trinity – composed of primordial violence, hatred, and enmity, which are to be regarded as a blind natural force; of the play of chance and probability within which the creative spirit is free to roam; and of its element of subordination, as an instrument of policy, which makes it subject to reason alone.

> The first of these three aspects mainly concerns the people; the second the commander and his army; the third the government. The passions that are to be kindled in war must already be inherent in the people; the scope which the play of courage and talent will enjoy in the realm of probability and chance depends on the particular character of the commander and the army; but the political aims are the business of government alone.

> These three tendencies are like different codes of law, deep-rooted in their subject and yet variable in their relationship to one another. A theory that ignores any one of them or seeks to fix an arbitrary relationship between them would conflict with reality to such an extent that for this reason alone it would be totally useless.

> Our task therefore is to develop a theory that maintains a balance between these three tendencies, like an object suspended between three magnets.

> What lines might best be followed to achieve this difficult task will be explored in the book on the theory of war (Book Two). At any rate, the preliminary concept

---

33 Clausewitz, 'Note of 1816–1818', 69.

of war which we have formulated casts a first ray of light on the basic structure of theory, and enables us to make an initial differentiation and identification of its major components.[34]

As we can see, although the people-army-government model has enjoyed the limelight in recent times, it is only in the second paragraph that Clausewitz links the core components of the Trinity to the people, army and government. In other translations, the link between the two is less well defined.[35] Anyway, it is clear that the people, army, and government components are there to provide context to the core idea; they do not act as an alternative.

Looking at the core concept then, Clausewitz begins by explaining that 'War is more than a chameleon'. By stating this he is explicitly moving away from discussion of the character of war, which will inevitably change its appearance over time. By applying the chameleon metaphor he is telling us to look beyond war's characteristics. For a real understanding of war's hidden traits we must study its nature. This he argues, is interactive, and is:

> Composed of primordial violence, hatred, and enmity, which are to be regarded as a blind natural force; of the play of chance and probability within which the creative spirit is free to roam; and of its element of subordination, as an instrument of policy, which makes it subject to reason alone.

He then explains that although the tendencies are 'deep-rooted', they are also 'variable'. He continues: 'A theory that ignores any one of them or seeks to fix an arbitrary relationship between them would conflict with reality to such an extent that for this reason alone it would be totally useless'.[36] By this, he is trying to convey the sense that when war takes place, it will always share these same proclivities, but each war will be different. Each will evolve along its own natural path.

Again, we must note that *On War* is a book designed to prepare soldiers with insights which they and the strategist can put into practice. Clausewitz thus cautions that theory 'maintains a balance between these three tendencies, like an object suspended between three magnets.' Any strategy that concentrates too much attention on one of the Trinity's component parts misunderstands the nature of war. We need to comprehend the whole: the way in which these tendencies interact and how each element impacts on the policies of the belligerents as their strategies are pulled and stretched, altered by real war characteristics. This he tells us is the 'first ray of light on the basic structure of theory'.

---

34   Clausewitz, *On War*, 101.

35   While the Paret/Howard translation suggests that the three elements 'mainly' correspond to the three sections of society, in the Graham translation, 'mainly' has been replaced by 'more', which has less direct connotations.

36   Clausewitz, *On War*, 101.

Of course, as was highlighted already, there is some controversy surrounding the Howard/Paret translation which is used above. Firstly, in terms of the choice of the word 'Trinity', it has been argued that Clausewitz was simply describing a three-part model. However, the connotation with religion is hard to ignore. Christopher Bassford argues that Clausewitz was a man of the Enlightenment and therefore rejected mysticism. This does not mirror the other influences on his work; as highlighted previously, Clausewitz actually rejected much of the rationalism of the Enlightenment. Additionally, we know that Clausewitz's relatives had positions within the Church; the imagery of the word would not have been lost on him. Directly linking this three part theory with religion gives it an aura of mysticism, a numinous quality which suggests a phenomenon slightly beyond our intellectual ken. The nature of war as elucidated by the Trinity is complex and unpredictable, even unfathomable.

This leads us to another point. As highlighted in the previous chapter, Howard has flagged problems with his and Paret's translation which may have impacted on the true meaning of the concept. When their translation is measured against earlier text, the Trinity takes on a different appearance which seems more in keeping with Clausewitz's intended meaning. In the Graham translation, the most widely known English alternative, the translation conveys the sense that 'reason' is just one part of the model. In the Graham translation the Trinity is composed of:

> The original violence of its elements, hatred and animosity, which may be looked upon as blind instinct; of the play of probabilities and chance, which make it a free activity of the soul; and of the subordinate nature of a political instrument, by which it belongs purely to the reason.[37]

Reason is one of three interrelated tendencies, and it appears to fit more naturally with Clausewitz's ideas regarding the critical influence of moral and neutral factors. While it may transpire that policy is a dominant component, war is certainly not subject to 'reason alone'.

The Trinity's First Component: Hostility

The first element in Clausewitz's theory – passion, hatred and enmity – plays a destructive but fundamental role in all wars and is often viewed as their dominant tendency. This is certainly the case in terms of the wars in Croatia and Bosnia-Herzegovina during the 1990s. Violence and animosity are thought to be largely responsible for the scale and brutality of the fighting. Yet, as Bassford points out, 'Clausewitz is not talking about physical violence', he is discussing 'violence' as a motivating force that can change the course of war, or even drive it on when

---

37   Clausewitz, *On War*, trans. Graham (ed.) (1873).

it no longer displays any rational purpose.[38] As Clausewitz describes it, the first component in his three-part model is 'composed of primordial violence, hatred, and enmity, which are to be regarded as a blind natural force'.[39] It is one of the core elements which drive war's natural tendency towards extremes.

Again in Chapter One, Book One; Clausewitz discusses the impact of hostility, concluding that there are two types of it. Firstly, there are 'hostile feelings', which are often present prior to war taking place; for example, in a shared cultural memory passed down from earlier generations. As Clausewitz explains in section eleven of Chapter One, Book One, 'Between two peoples and two states there can be such tensions, such a mass of inflammable material, that the slightest quarrel can produce a wholly disproportionate effect – a real explosion'.[40] Despite this, Clausewitz is quite clear in his position that 'hostile feelings' do not necessarily need to exist prior to war beginning. Highlighting the same position, Echevarria notes that 'the populations of two warring states need not have any basic animosity toward one another for the states to be at war'.[41] There have been plenty of examples of wars where the populations of the belligerents have enjoyed good relations prior to conflict.

Before the 1991 Gulf War, there was little animosity between the Iraqi people and the populations of Iraq's adversaries. Turning to a Balkan example once more, one thinks of the friendly relationship between Serbs and Slovenes before Slovenia's successful war of secession from Yugoslavia in 1991. One also thinks of the UK's involvement in the targeting of Serbia in 2000. The two states had traditionally been allies and there was little in the way of hostile feelings emanating from the Serbs; equally, there was little in the way of animosity emanating from the UK populace. In tribute to their one-time solidarity, a picture commemorating their alliance hangs in the Foreign Office in London. For many Serbs, the UK's role in the raids on Serbia provoked bewilderment rather than hostility. As Clausewitz explains, 'Even the most savage, almost instinctive, passion of hatred cannot be conceived as existing without hostile intent; but hostile intentions are often unaccompanied by any sort of hostile feelings'.[42]

The second type of hostility – 'hostile intentions' – on the other hand is, at least for one side, always present. Thus, hostile intentions may often be unaccompanied by any sort of hostile feelings, they may simply be a component of policy. However, the hostile intentions of one side inevitably provoke a cause-and-effect relationship that is reciprocal and which can push each belligerent to extremes, exacerbating hostility still further. In addition to this, there is also the possibility that 'hostile

---

38   Christopher Bassford, 'Tip-Toe Through the Trinity, or The Strange Persistence of Trinitarian Warfare', www.Clausewitz.com/CWZHOME/TRINITY/TRINITY8.htm, (2007), 10 (accessed 2 January 2009).

39   Clausewitz, *On War*, 101.

40   Clausewitz, *On War*, 90.

41   Echevarria, *Clausewitz and Contemporary War*, 72

42   Clausewitz, *On War*, 84. See also: Kalyvas, *The Logic of Violence in Civil Wars*.

intentions' may actually foster 'hostile feelings' as a matter of strategic calculation. This may be a calculated decision that provokes resentment in another people, or equally, it may be designed to stoke the passions and hostility of one's own people in preparation for the depravations of war to follow. This is a charge laid at the feet of the competing elites within Yugoslavia's fragmenting political structure.[43]

Although hostility – passion, hatred and enmity – may be generated for myriad reasons, and although Clausewitz believed that every war would display unique fluctuations of 'violence' driven by hostility, it was, nevertheless, an element which is universal to all wars, across all time periods. Clausewitz himself explores its escalatory properties in the opening chapter of his tome, laying the foundations for future misinterpretations, but also placing physical contest at the heart of his theory. To understand war holistically, Clausewitz thus uses 'hostility' as one of the elements in his three-part theory of war. Of course, it is axiomatic that hostility can exist independently of conflict; however, it is equally evident that no theory of war which aims to elucidate a general paradigm can leave it out of the equation. In particular, the cause-effect relationship which is at the heart of Clausewitz's understanding of conflict, and which exacerbates hostility, is conflated further by the interaction of the two further tendencies of the Trinity.

Although the extent of this interplay will be explored in the concluding chapter, Chapter 4 deals primarily with gauging the extent to which 'passion, hatred and enmity' impacted on the nature of the wars of former Yugoslavia between 1991 and 1995. In his discussion of hostility Clausewitz compares the passions inherent in savage warfare with those in that of 'civilized' people, yet he comes to the conclusion that in all wars, hostility, manifested by the passions, hatred and enmity released in war, is a universal element. Even 'civilized' peoples will be shaped by these forces. Conceptualizing war without factoring in its fundamental features would delimit the purpose of *On War*, undermining the cogency of its ideas as a consequence. And that, as Clausewitz tells us, 'would be a kind of war by algebra'.[44] When Clausewitz expresses his opinion that there is not a set formula to gauge the impact of hostility he is reflecting his view that war is unpredictable and that each will inevitably display its own peculiarities.

The Trinity's Second Component: Chance and Uncertainty

In section twenty of Chapter One, Book One, Clausewitz tells us that 'Only one more element is needed to make war a gamble – chance … No other human activity is so continuously or universally bound up with chance'. He continues:

---

43   See V. Gagnon, 'Ethnic Nationalism and International Conflict: The Case of Serbia', *International Security* (Winter, 1994–95), 130–66; and Eric Gordy, *The Culture of Power in Serbia: Nationalism and the Destructions of Alternatives* (Pennsylvania: Pennsylvania State University Press, 1999).

44   Clausewitz, *On War*, 84.

'And through the element chance, guesswork and luck come to play a great part in war'.[45] How should the reader of *On War* interpret Clausewitz's inclusion of chance in his theory of war? Why is war so 'universally' intertwined with the pervasive tendencies of chance? In fact, what does Clausewitz actually mean by 'chance'?

Clausewitz's introduction of chance as an integral component of war's nature is a direct refutation of what he believes are perfunctory and misleading elements found in the theories of his contemporaries. Their theories sought to uncover inexorable laws which lay hidden, waiting to be discovered. Although Jomini concedes the existence of chance, he dismisses its importance because he argues that the identification of set strategic rules can overcome its pervasiveness. He was in good company. According to the Scottish Enlightenment's, David Hume, 'chance is nothing but a secret and conceal'd cause'.[46] Reflecting on this notion, Ian Hacking reformulates the Enlightenment thinkers' belief: the world 'might look haphazard, but only because we do not know the inevitable workings of its inner springs'.[47] Clausewitz's contemporaries sought to find these rules; their focus on probabilities rested on a lack of knowledge which could be replaced by hard facts as and when these became evident. Clausewitz is critical of these writers because they seek fixed rules where none exist. Clausewitz explains, 'In short, absolute, so-called mathematical, factors never find a firm basis in military calculations'.[48] 'Military activity is never directed against material force alone; it is always aimed simultaneously at the moral factors which give it life, and the two cannot be separated'.[49]

As in other aspects of his thinking, Clausewitz was not the first to explicate the importance of stochastic principles.[50] Although Clausewitz was a product of the Enlightenment, he tapped into the burgeoning ideas of the Counter-Enlightenment and we know that he was influenced by the ideas of German Romanticism. Although he never fully embraced this movement, he was influenced by its ideas regarding the place and influence of psychological factors; a response to the increasingly statistical theorizing which dominated military and scholastic writing.[51] In many ways Clausewitz was stating the obvious, even banal – chance is everywhere. And this is the very point he was making and which had been overlooked by earlier thinkers. In the Clausewitzian sense, armies are not automatons, and war is not linear. Like the rest of life, strategy and war are prone to the pervasive tricks of

---

45   Clausewitz, *On War*, 96.

46   David Hume, *A Treatise of Human Nature* (London: J.M. Dent & Sons, 1739), L.A. Selby-Bigge (ed.) (Oxford, 1888), 130, cited in Ian Hacking, *The Taming of Chance* (Cambridge: Cambridge University Press, 1990), 13.

47   Hacking, Ibid., 1

48   Clausewitz, *On War*, 97.

49   Clausewitz, *On War*, 137.

50   Clausewitz used the work of earlier writers as a foundation for his own work, Machiavelli and Thucydides in particular. Machiavelli discusses the subject in *The Prince*, vol. 1, 89–92.

51   Gat, 189.

chance. Clausewitz was introducing the real world, a reminder that war doesn't take place in a separate realm, subject to alternative laws. War is a product of the social, not physical world.

In itself this inclusion is important; it reminds us of the peculiarities of life. However, Clausewitz goes further. He states: 'No other human activity [as war] is so continuously or universally bound up with chance'.[52] The answer, as always, lies in war's interactive form. Clausewitz exclaims, 'war is not waged against an abstract enemy, but against a real one who must always be kept in mind'.[53] War 'consists of a continuous interaction of opposites'.[54] It is within this interaction that chance reigns. It is a normal and indelible component of the nature of war, and it pervades every element of conflict. As Colin Gray remarks, 'the would-be rational and prudent defence planner lives in a world of uncertainty, chance does not quite rule but it is always a player'.[55] Although chance may not rule supreme in the minds of the strategist, one should not underestimate the potential of chance to bedevil the commander and to derail the best laid plans. Clausewitz explains further:

> From the very start there is an interplay of possibilities, probabilities, good luck and bad that weaves its way throughout the length and breadth of the tapestry. In the whole range of human activities, war most resembles a game of cards.[56]

Ian Hacking explains the inner workings of this connectivity and its effects by reference to 'the long standing idea of intersecting causal lines.' He states:

> Suppose that you and I meet 'by chance' at the market. There may be a causal story of why I am at the market at ten past nine in the morning, choosing cantaloupes. A different but equally causal account will explain why you are there at that time, picking your peaches. Because the two sets of causes together entail that we will cross paths at 9.10, there was nothing 'determined' about our meeting. We call it chance, but not because the event was uncaused. Chance is a mere seeming, the result of intersecting causal lines.[57]

Chance is magnified by the interaction of opposites. Although chance is everywhere, the intersection of lines to which Hacking refers complicates action and produces uncertainty; and the more intersecting lines, the more uncertainty.

---

52   Clausewitz, *On War*, 96.
53   Ibid., 187.
54   Ibid., 136.
55   Colin S. Gray, *Modern Strategy* (Oxford: Oxford University Press, 1999), 41; see also: Katherine L. Herbig, 'Chance and Uncertainty in *On War*', in Michael Handel (ed.), *Clausewitz and Modern Strategy*, 95–115.
56   Clausewitz, *On War*, 97.
57   Hacking, *The Taming of Chance*, 12.

In its plainest form then, 'chance' refers to the unforeseen and reflects the very interactivity which is at the heart of the Trinity. From the moment that war commences, a dynamic relationship begins which is constantly developing its inimitable form; constantly in motion, changing as the motivations and passions of the belligerents fluctuate; and conflating and expanding further as the interaction converges with the other forces in a reciprocal interplay with hostility and policy. Chance interferes with every aspect of war, from resistance encountered in the military machine to the difficulties of working with allies, as is exemplified by Clausewitz's famous concept of 'friction'. Under such conditions, one can never entirely be sure what our opponent is planning.[58] Uncertainty pervades the decision-making process as a result.[59] 'War is the realm of uncertainty; three quarters of the factors on which action in war is based are wrapped in a fog of greater or lesser uncertainty'.[60] He reiterates his point once more:

> War is the realm of chance. No other human activity gives it greater scope: no other has such incessant and varied dealings with this intruder. Chance makes everything more uncertain and interferes with the whole course of events.[61]

It was this uncertainty which brought Clausewitz to conclude that military decisions should not be premised solely on intelligence. 'Many intelligence reports in war are contradictory; even more are false, and most are uncertain'.[62] His refusal to recognize the importance of intelligence to war over time has been linked to the false notion that Clausewitz propounded the theory of total war.[63] It has been claimed that Clausewitz was suggesting that power should trump intelligence. This was not what Clausewitz was saying at all. Rather his aversion to intelligence as an avatar of easy strategic success rests on his appreciation of the complicating

---

58   Despite this, Clausewitz dedicates part of Book Eight to this topic. Although he gives detailed principles on planning, these should be thought of as guidelines rather than fixed maxims. For an interesting contradistinction on the topic of planning in *On War*, see: Terrence M. Holmes, 'Planning versus Chaos in Clausewitz's *On War*', *The Journal of Strategic Studies*, 30(1) (February 2007), 129–51.

59   See: Stephen J. Cimbala, *Clausewitz and Chaos: Friction in War and Military Policy* (Westport, Connecticut: Praeger, 2001).

60   Clausewitz, *On War*, 117.

61   Clausewitz, *On War*, 117. While in Book Two, chance is the 'intruder', which 'interferes with the whole course of events', in the Trinity (and thus the finished Book One) chance represents uncertainty, but interlinked to his understanding of psychology and his adherence to the notion of genius, it also displays a positive side: chance provides the arena where 'the creative spirit is free to roam'. This point is also made by: Herbig, Chance and Uncertainty', 95–116.

62   Clausewitz, *On War*, 136.

63   David Kahn, 'Clausewitz and Intelligence', Michael Handel (ed.), *Clausewitz and Modern Strategy*, 116–26.

effects of the interaction of chance and uncertainty. In a complex milieu such as war, perfect intelligence is a misnomer. He explains:

> Since all information and assumptions are open to doubt, and with chance at work everywhere, the commander continually finds that things are not as he expected ... During an operation decisions have usually to be made at once: there may be no time to review the situation or even to think it through. Usually, of course, new information and revaluation are not enough to make us give up our intentions: they only call them in question. We now know more, but this makes us more, not less uncertain. The latest reports do not arrive all at once: they merely trickle in. They constantly impinge on our decisions, and our mind must be permanently armed, so to speak, to deal with them.[64]

In *On War* Clausewitz relates this problem of war to the ability of the commander to survey the possible probabilities, gauge the possible opportunities, pitfalls, and outcomes, and act decisively. It was this aspect of war which required the quality of 'genius', and he devotes the whole of Chapter Three of Book One to this topic, where he lauds the virtues of the commander able to deal with the complexities of real conflict. It is only the genius who is capable of recognizing the opportunities 'that the mind would ordinarily miss or would perceive only after long study and reflection'.

In Clausewitz's Trinity, this intuitive side of war's nature is dialectically opposed to the rationality which underpins policy. However, through the complex interaction of purpose and hostility, chance mingles and converges into every pore of war's nature. As this phenomenon is formed through the reciprocal interplay of opposites which Clausewitz describes, it creates and then exacerbates the problems which all wars exhibit. However, as Terrence Holmes has demonstrated, chance also creates opportunities for the commander able to display acuity and nerve.[65] Clausewitz's inclusion of this component does however mean that it cannot be tempered. The interplay of chance and probabilities reflects war's uncertainty and the requirement of those waging it to work in 'the twilight, which, like a fog or moonlight, often tends to make things seem grotesque and larger than they really are'.[66] Acting on 'probabilities' rather than reliable information, the commander, and by extension the statesman, must demonstrate skill and dexterity if they are to succeed in their venture.[67]

---

64   *On War*, 117.

65   Terrence Holmes, 'Planning versus Chaos in Clausewitz's *On War*', *The Journal of Strategic Studies*, 30(1) (February, 2007), 129–51.

66   Ibid., 140.

67   In contrast to other important war theorists, Clausewitz's view of uncertainty is by far the most pessimistic. Sun Tzu argues that Intelligence can overcome uncertainty. Although Jomini also acknowledges the place of chance, he argues that scientific principles can relegate the dangers of uncertainty. Mao Tse-Tung, himself a disciple of Clausewitz,

The Trinity's Third Component: War's Subordination to Policy

The third element within the Trinity – policy/reason – requires careful examination; not least because it has been subsumed within the concept rather than enjoying preeminent status pertaining to war's subordination to policy. This has undoubtedly caused frequent problems for readers attempting to grasp Clausewitz's true message. From as early as 1827, and explicit throughout Chapter One, Book One, is the idea that war 'is nothing but a continuation of policy with other means'.[68] Yet, when he conceptualizes the Trinity, 'policy' has an equal role with the other tendencies in the concept – hostility and chance. War as an act of policy has been superseded by the Trinity – his general theory of war.

Clear comprehension of Clausewitz's intended argument has unquestionably been hampered by the confusion surrounding his original word choice – he uses the word *Politik*, which has no direct English translation. Readers of the English translations are thus confronted with a choice between 'politics' or 'policy', and the two are often used interchangeably. As highlighted in the previous chapter, in their translation of *On War*, Paret and Howard preferred to use 'policy' more frequently than politics. The decision to do so stemmed from their feeling that 'politics' was too closely associated with the murky world of domestic politics and a general malaise of politicians; it thus held too many negative connotations. Policy, on the other hand, was something that states pursued to maintain national interests on the grand-stage of international relations.

Yet how should the reader interpret these words? Moreover, how do we separate politics and policy in the English translations of *On War*? Thankfully, there has been work to highlight the critical but often overlooked differences between the two. As Bassford rightly points out, while politics and policy are both concerned with power, they exemplify different forms. On the one hand, 'politics' is the 'process by which power is distributed in any society', an interaction of competing ideas. In turn, Bassford continues, policy is 'the rational and one-sided subcomponent of politics, the reasoned purposes and actions of each of the various individual actors in the political struggle.'[69] For him, the difference again lies in interactivity:

> Politics is a multi-lateral phenomenon, whereas policy is the unilateral subcomponent thereof. My ally, myself, and my enemy are all bound up together in politics, but we each have our own policies.[70]

---

also argues that uncertainty and chance can be overcome by careful planning. Rather than probabilities, there are 'relative certainties'. A review and analysis of the contrasting ideas can be found in: Michael I. Handel, *Masters of War*, third edition (London: Frank Cass, 2001), 215–53.

68   Clausewitz, 'Note of 1827', 77.
69   Bassford, 'Tip-Toe Through The Trinity', 13.
70   Ibid.

When inferring Clausewitz's meaning then, we can tell that on a multilateral level, war remains a continuation of politics; indeed, in the Howard/Paret edition this is how they translate it:

> We see, therefore, that war is not merely an act of policy but a true political instrument, a continuation of political intercourse, carried on with other means. What remains peculiar to war is simply the peculiar nature of its means.[71]

Conversely, war as a 'continuation of policy' is unilateral – it reflects the decision to use military force as a means of reaching a political objective. Although in one sense the differences between politics and policy are axiomatic, in *On War* clarity is most certainly lacking. Of course, Clausewitz's use of the word *Politik* covers both politics and policy, but it is not always easy to identify whether he is discussing war in the bi-lateral/multi-lateral sense, or unilaterally – a common complaint articulated by Clausewitz's critics.

Returning our attention to the Trinity, in Clausewitz's conceptualization, policy is one element of a three-part model. War is a multilateral, political activity, while policy is only one of three competing tendencies within it. According to this line of thinking, although unilateral policy may instigate war, as soon as it does so it becomes a multilateral, reciprocal activity. The policy of the aggressor meets with the other forces of the Trinity and, not least, with the competing policy of one's opponent.

As we know already, Clausewitz tells us of his intention 'to develop a theory that maintains a balance between these three tendencies, like an object suspended between three magnates.' Basically, what he is telling us is that war is complex and unpredictable – it is in fact, to be precise, non-linear. The Trinity is intended to highlight this; it is, as he argues, 'a first ray of light on the basic structure of theory'. Hew Strachan reminds us: 'The three elements jostle for primacy, Clausewitz likening them to magnets in their capacity not only to attract but also to repel. The outcome is therefore almost infinitely variable.'[72] Through it the reader is presented with real war; it does not follow a 'linear' path from policy to peace treaty. As soon as it begins, reciprocity produces a cause-and-effect relationship which is non-linear in its outcome, where the policies of each belligerent are bashed around by hostility and chance. An appreciation of this interaction is the starting point of good strategy, precisely because it explicates the complex nature of war. The policy-maker and strategist must be aware that war is a constantly moving phenomenon.

As a consequence of this, strategy must therefore reflect the changes that take place. Failing to fully understand the labile nature of conflict is apt to end in stalemate, or worse, defeat. The Trinity tries to explain this interaction, especially the way in which policy needs to adapt in consequence to interplay

---

with the other elements in the model. As this book attests, the Trinity reflects a valid and illuminating insight into the nature of war which stands the test of time. The continual interaction and the complexity which the Trinity explains is a vital place to found our strategic knowledge, and is in sharp contradistinction to those who believe that war can be rid of complexity and unpredictability. As the following chapters demonstrate, however, this interaction does not completely subsume the centrality of the political (policy) in Clausewitz's treatise. Moreover, reconnecting this element and realigning the Trinity to account for this provides a better understanding of Clausewitzian theory and the nature of war in our modern period.

Interpreting the Clausewitzian Trinity:
A Framework of War in the Modern World?

In the light of Clausewitz's seemingly universal model, what direction should future research follow? The obvious answer is to pursue a Clausewitzian analysis of war in the modern world; using Clausewitzian theory as the bedrock of strategic calculation. Certainly, it would be less than prudent to discard his ideas as irrelevant to our own period. Clausewitz has bequeathed a concept that appears to apply to any period; this book validates the Trinity by grounding it to the experience of real war. If the Trinity is to retain its status it is important both to test the strength of his ideas against contemporary examples, and, if appropriate, refine them to enable a modern analysis of war. This seems a particularly pertinent route when one surveys the outcomes of this most recent round of interpretative analysis. As this book illustrates, grounding the Trinity to experience the Trinity illuminates the predominant position of policy within the Triad, a notion rejected in the current literature.

As we know, Bassford, Echevarria, and Herberg-Rothe all situate the Trinity at the heart of Clausewitzian theory. Yet, while these writers have accurately illuminated Clausewitz's 'core' concept over the people, army, government model, there is still a requirement to test this idea against empirical evidence. By carefully analyzing *On War* they build a compelling case that his Trinity is central to his entire tome, and moreover, that his insights are transferable to our own period. Nevertheless, while assessing and interpreting *On War* as a core and primary text has helped explicate Clausewitz's 'real' Trinity, by exploring his ideas against experience we can gain a better understanding of the concept and its place within the rest of Clausewitzian theory. Recent Clausewitzian scholarship should be applauded; it has reignited interest and accessibility into Clausewitz's ideas. However, its analyses have also highlighted an unmistakable tension at the heart of his theory. In their determination to prove the legitimacy of the Trinity, Clausewitz's advocates have clashed with his insistence that war is instrumental, 'a continuation of policy'.

As we know already, Clausewitz is perhaps most famous for the above dictum. Though Gray has correctly observed that the formula has been held captive to folly, this Clausewitzian idea has formed one of the cornerstones of strategic theory.[73] It was the link between politics and war which ensured Clausewitz's ideas would not be erased by the vagaries of the historical process. It has been this relationship which has sustained enduring interest in his writings. Nevertheless, despite Clausewitz's various assertions that 'war is the continuation of policy', when he articulates his Trinity, policy is subsumed by the competing elements in the formula. Supporters of the Trinity mean to reflect his insistence that anyone seeking 'to fix an arbitrary relationship between them would conflict with reality to such an extent that for this reason alone it would be totally useless'.[74] This helps attune readers to the fact that war is ultimately a very unpredictable, voluble activity. Conveyed in this way, the Trinity is instructive to any group engaged in conflict; it is not restricted to the state centrism which so underpinned comprehension of international relations throughout the last century. The problem arises, however, when Clausewitz explains that: 'Our task therefore is to develop a theory that maintains a balance between these three tendencies, like an object suspended between three magnets.'[75] This clashes with Clausewitz's other more famous dictum that war is the continuation of policy, and thus instrumental – the foundation of modern Western strategic theory.

The Trinity displays a unified theory where its three tendencies are part of a whole, displaying unpredictability and complexity. By the end of Chapter One, Book One of *On War*, war as an act of 'policy' has been subsumed by the Trinity. Thus, within the Trinitarian formula, 'policy' is equal to 'passion' and 'chance'. Notwithstanding the genuinely insightful exploration of interaction and interplay that Clausewitz conjures up by way of his model, is he right to give the three elements of the Trinity equal weight? Although the idea that war is 'simply the continuation of policy' has been used to bash Clausewitz, in their haste to defend his 'strategic universe' have his proponents overlooked a main pillar of his argument? Did Clausewitz intend the Trinity to eclipse his clear political message?

This is an important point, one which Clausewitz never resolved before his early death. The voluminous extent of literature based on Clausewitzian studies has also failed to adequately balance this tension. In one rare example, Gallie does seem to identify the problem, the 'contradiction' as he puts it, but he quickly skirts away from it to take up the closely connected case of limited versus absolute war. It is a question that deserves further attention. If one element is more important than the other, then there may be a need to refine his theory. His unfinished symphony, which Echevarria claims should be left well alone, may require to be finished.

---

73 Gray, *Modern Strategy*, 30.

74 Clausewitz, *On War*, 101.

75 Ibid.

Let us examine this point in more detail. Although a supporter of the Trinity as central to Clausewitz's opus, Echevarria argues that:

> The Trinity conveys the sense that none of the tendencies of the nature of war is *a priori* more influential in determining the shape and course of actual conflict than any other. Thus to single out policy or politics, as the central element of war's nature is to distort the intrinsic balance implied by the mere concept of the Trinity itself, and ultimately to compromise its dynamism.[76]

It is a view supported by Herberg-Rothe, who notes that: A complete reduction of war to an instrument of policy would be a contradiction even within Chapter I, as it would conflict with the 'Wondrous Trinity'.[77] He continues, 'War as a continuation of policy by other means itself rests on a tension that can never be overcome.'[78]

The problem with the Trinitarian approach as espoused by Echevarria and Herberg-Rothe is that the Trinity usurps Clausewitz's other claim that war is an act of force to compel one's enemy to one's will. In the Trinity, war's reciprocal, multilateral (political) interplay makes it unpredictable and complex. When war in its multilateral sense takes place, the constant interaction between the three tendencies of the Trinity is paramount. In other words, war as an act of policy is of equal importance to passion and chance. This raises important points which have not been adequately addressed in the recent literature. For instance, the idea that 'policy' is firstly the cause of conflict, but is then subsumed by equally important tendencies needs to be questioned. Is policy really not dominant within the concept? If it is not, then the Trinity is a model that demonstrates complexity but it does not do much more than that. Is this really what Clausewitz intended? If war should be thought of as 'purely' Trinitarian then this delimits Clausewitz's other arguments. Although he never resolved this tension, he does inform us in his 'Authors note of 1827' that he has decided to revise his work. This is the product of two closely related insights. Firstly, war is a continuation of policy, and secondly, that it can be of two kinds: real and ideal. Clausewitz remarks:

> The distinction between the two kinds of war is a matter of actual fact. But no less practical is the importance of another point that must be absolutely clear, namely that war is nothing but the continuation of policy with other means. If this is firmly kept in mind throughout it will greatly facilitate the study of the subject and the whole will be easier to analyze.[79]

Of course, the conceptualization of the different components of his theory 'balancing between three magnets' helps invoke a sense of war's constant

---

76   Echevarria, *Clausewitz and Contemporary War*, 73.
77   Herberg-Rothe, *Clausewitz's Puzzle*, 115.
78   Ibid., 117.
79   Clausewitz, 'Authors Note of 1827', 77.

interactivity. However, does this model also mask the extent to which one of the elements of the Trinity – policy – is more important than the other components? Can war be Trinitarian and instrumental at the same time? Let us dwell on this point. We know that Clausewitz's model depicts the reciprocal and therefore multilateral nature of war, and policy is one of three equal parts of that theory. This highlights the extent to which policy is bashed around, and its course altered as real war features – hostility and chance – deflect and erode the original purpose (policy) that brought war into being. However, we must remember that Clausewitz was writing as a strategist; he finds the three-part nature of war in his search to find the best way of linking purpose and means. Echevarria argues that Clausewitz's Trinity 'negates the notion of the primary of policy; it renders policy as purpose, and holds it *a priori* just as important as chance and hostility'.[80] However, we would do well to remember that both policy and purpose do not exist independently from the political entity that initiated it. War, Clausewitz tells his readers, 'cannot be divorced from political life'.[81]

Whether or not Clausewitz's appreciation of war as an act of politics developed late in his work as envisaged by Azar Gat is a point of debate. However, if his conceptualization of war as an act of politics did shatter his earlier conception of theory, this does not invalidate the central point which he gleaned from his study and experience. Not only was Clausewitz increasingly cognizant that war was indelibly linked to politics, this also affected the strategic principles he was attempting to clarify. In his author's note of 1827, he advises us that he has decided to rework *On War* because of his realization that it could be of two kinds, real and absolute. The reason for his real/absolute war dichotomy relates directly to the influence of politics and the friction of real life. The character and levels of violence attributed to a particular war are linked to the causes of war, and strategy must take this into account as it attempts to achieve the ends of policy. In a letter to his friend at the Prussian General Staff, Major von Roedor, Clausewitz explains the hollowness of devising plans without taking into account the character of the belligerents or the reasons why they may want to go to war. He remarks:

> *War is nothing but the continuation of political efforts by other means.* In my view all of strategy rests on this idea, and I believe that whoever refuses to recognize that this must be so does not yet fully understand what really matters. It is this principle that makes the entire history of war comprehensible, which in its absence remains full of the greatest absurdities.[82]

Continuing in the same vein, Clausewitz comments:

---

80   Echevarria, *Clausewitz and Contemporary War*, 95.

81   Clausewitz, *On War*, 731.

82   Clausewitz, 'Two Letters on Strategy', in Peter Paret, *Understanding War: Essays on Clausewitz and the History of Military Power* (Princeton: Princeton University Press, 1992), 127.

> I must insist that the military goals of both sides are stated whenever a strategic plan is drawn up. For the most part these goals arise out of the political relations of the two antagonists to each other, and to other states that may be involved … No one tries to push his argument back to the origins of the war that is to be fought, to its true motive, to the one and only point where the logical development and conclusion of the military operations can alone originate.[83]

As one of Clausewitz's contemporaries, Otto August Ruhle von Lilienstern, argued, 'There is a Why? and a What For?, a purpose and a cause, at the bottom of every war.'[84] Policy is thus firmly linked to the political community which enacted it, a position also propounded in *On War*. Clausewitz notes, 'Policy, of course, is nothing in itself; it is simply the trustee … And here we can only treat policy as representative of all interests of the community'.[85] This conviction sits uneasily with the alternative view that policy is subverted by war's nature. In the Trinity it seems that Clausewitz chose to elide his own political argument. As noted in *On War*:

> Rather than comparing it (war) to art we could more accurately compare it to commerce, which is also a conflict of human interests and activities; and it is still closer to politics … Politics, moreover, is the womb in which war develops – where its outlines already exist in their hidden rudimentary form, like the characteristics of living creatures in their embryos.[86]

Within the Trinity, 'policy' is limited by the internal factors of war's nature; it is forced to change its form when it interacts with hostility and chance, and the policy of one's opponent, which inevitably clashes with one's own. By highlighting that the tendencies within his concept are 'deep-rooted' and 'variable', Clausewitz alerts us to the extent to which they interact through war's non-linear traits. But did he intend them to be equals? However much the policies of competing belligerents are affected by the dynamism of the Trinity's core elements, like strategy, policy will alter and feed its way back into the polity from which it came. An adjusted, updated policy is then fed back into the cauldron that is the Trinity. It may be altered by the competing tendencies, but whether it is good or bad policy, it can still change as the elements of the Trinity interplay with each other. War is Trinitarian, but just as the Trinity shapes policy, so too is the Trinity shaped by the political purpose that brought it into being. Should we believe that the link is severed when war begins? Clausewitz asserts:

---

83   Clausewitz, 'Two Letters on Strategy', 129.

84   Otto August Ruhle von Lilienstern, *Handuch fur den Offizer zur Belehrung im Frieden und zum Gebrauch im Felde*, Vol. II (Berlin: G. Reimer, 1817), 8; cited in Heuser, *Reading Clausewitz*, 30.

85   Clausewitz, *On War*, 733.

86   Clausewitz, *On War*, 173.

That the political view should wholly cease to count on the outbreak of war is hardly conceivable ... In fact, as we have said, [wars] are nothing but expressions of policy itself. Subordinating the political point of view to the military would be absurd, for it is policy that has created war. Policy is the guiding intelligence and war only the instrument, not vice versa.[87]

He continues:

Once again: war is an instrument of policy. It must necessarily bear the character of policy and measure it by its standards. The conduct of war, in its great outlines, is therefore policy itself, which takes up the sword in place of the pen, but does not on that account cease to think according to its own laws.[88]

He reiterates this line of thinking again in his revised Chapter One, Book One: 'The smaller the sacrifice we demand of our adversary, the smaller we can expect his efforts to be to deny us. The smaller his effort, the smaller ours can be.'[89] Although Echevarria rejects policy's primary status, he concedes that 'the objective and subjective natures of war are not separate phenomena but rather aspects of the same phenomenon'.[90] Of course, when arguing that political purpose underpins all wars, Clausewitz notes that the political purpose is not 'a tyrant':[91]

It must adapt itself to its chosen means, a process which can radically change it; yet the political aim remains the first consideration. Policy, then, will permeate all military operations, and so far as their violent nature will admit, it will have a continuous influence upon them.

Commenting on the last sentence of the above extract, Herberg-Rothe has claimed that policy is perverted by war and thus holds equal weight to the other tendencies in the Trinity. He argues: I would like to suggest that this formulation is where Clausewitz provides his best description both of the influence of policy on warfare and the limits of its importance.'[92] However, it could just as easily be interpreted as an acknowledgement that policy is constantly evolving as it interacts with war's violent and complex, non-linear nature. As Clausewitz put it, 'it (policy) will have a continuous influence'. Elsewhere in *On War*, Clausewitz warns his readers of the fallacy that political relationships are extinguished when war begins:

---

87   Clausewitz, *On War*, 733.
88   Clausewitz, *On War*, 737.
89   Clausewitz, *On War*, 98–9.
90   Echevarria, *Contemporary War* 77.
91   Clausewitz, *On War*, 98–9.
92   Herberg-Rothe, *Clausewitz's Puzzle*, 114.

> … It is apt to be assumed that war suspends that intercourse and replaces it by a wholly different condition, ruled by no law but its own.

> We maintain, on the contrary, that war is simply a continuation of political intercourse, with the addition of other means … war in itself does not suspend political intercourse or change it into something entirely different. In essentials, that intercourse continues, irrespective of the means it employs.[93]

In a recent exposition of Clausewitz's ideas, Hugh Smith argues war 'does not replace policy or relegate it to a lesser position … It is always a continuation "with" (mit) not a substitution "by" (durch) other means.'[94] Of course, Clausewitz concedes that 'The original political objects can greatly alter during the war and may finally change entirely since they are influenced by events and their probable consequences.'[95] As highlighted above, the political aim is not a 'tyrant', 'It must adapt itself to its chosen means, a process which can radically change it.'[96] Could this be what the Trinity is supposed to convey: policy is shaped and moulded by the unpredictability and complexity of war, often to the extent that the original motive is lost, but policy remains on a higher plane? Not only does policy shape war as much if not more than hostility and chance, it is policy which eventually results in victory or defeat. Clausewitz exclaims: 'The main lines along which military events progress, and to which they are restricted, are political lines that continue throughout the war into the subsequent peace. How could it be otherwise?'[97] This attitude is in sharp contradistinction to the supposition that the Trinity conveys total equality of opposites. Unlike hostility and chance, 'policy' is devised by people to achieve political goals through military methods. Moreover, as hostility and chance effect the course of the war, leaders adjust policy in response.

The Consequences for Theory?

There is a danger that the Trinitarian analysis, for all of its genuine insight, undermines its intended utility. While the Trinity elucidates a complicated phenomenon, if we pause to consider its purpose we can extrapolate further clues that the Trinity is incomplete in Clausewitz's thinking. Jon Sumida's recent study *Decoding Clausewitz* suggests the Trinity is merely a reference point and should not warrant the attention it is beginning to receive. Sumida's criticism stems from his own argument that the real message of *On War* is that defence is

---

93   Clausewitz, *On War*, 731.
94   Hugh Smith, *On Clausewitz: A Study of Military and Political Ideas* (Basingstoke: Palgrave MacMillan, 2004), 110.
95   Clausewitz, *On War*, 104.
96   Clausewitz, *On War*, 98.
97   Clausewitz, *On War*, 731.

the stronger form of war, offence the weaker. This doesn't actually mean that the Trinity is not important to current analysis. That Clausewitz meant the Trinity as a reference point does not delimit its modern validity. Clausewitz would not be the first to uncover a significant truth while trying to solve another riddle. Sumida's point is instructive, however. As he argues, Clausewitz's treatise is chiefly about empowering soldiers with lessons of strategy. True, Clausewitz was adamant that his strategic principles were not prescriptive, but he spent a great deal of time articulating strategic insights. From Book Three 'On Strategy in General', through to Book Eight, 'War Plans'. Clausewitz explores the ways in which soldiers can best overcome war's erratic nature – a point also made by Terrence Holmes.[98] While the study of Clausewitz has been greatly assisted by Beyerchen's linkages of war to non-linear theory, both Sumida and Holmes contend that the focus on the non-linear aspects of Clausewitz's work has eclipsed its normative elements, which prepare soldiers in how to overcome non-linearity. While this falls short of prescription, it does instil students with the conceptual tools needed to fight wars. *On War* is a strategy manual, and while Clausewitz does set out to understand and then explain the nature of war, he does so in order that the soldier is enabled with the knowledge and intellectual dexterity which is required to overcome the fluidity of real war. In the remainder of his work he introduces students both to the conduct of operations and also to the best and most profitable routes to strategic success in war's complex milieu. In other words, *On War* provides a genesis for strategic action.

Strategy is the bridge between war and policy, it should adapt and evolve to the changing circumstances of actual combat, shadowing policy as it changes throughout the course of conflict. When one analyses the purpose of strategy against the Trinity as articulated by Clausewitz, it would seem that strategy should attend to the alterations of the competing elements. In fact, as a reference point this initially seems to make sense. The Trinity expounds the idea that war is unpredictable at best – having a good grasp of the ways in which it can fluctuate is critical. The problem here is that strategy is only strategy when linked to the political purpose which put it into action. Strategy should follow policy's designs, not be caught in the unending struggle between the three points of the triangle. The difference may be subtle, but it is critical if good strategy is to follow.

Although the explication of the Trinity takes place in the one chapter that commentators consider finished, should one paragraph, despite its utility, shape future discourse? As an explanatory model is the Trinity flawed? It is a point which deserves further study. As a result of this tension, this book explores two closely related issues. Firstly, can the Clausewitzian Trinity retain its status

---

98   See: Terrence M. Holmes, 'Planning versus Chaos in Clausewitz's *On War*', *The Journal of Strategic Studies*, 30(1) (February 2007), 129–51. Holmes' article chiefly tackles the assumption that 'chance' is more important than what Clausewitz had to say about planning. However, his argument that too much emphasis is placed on war's non-linear elements at the expense of thoughtful analysis of his arguments is of interest here too.

as an analytical model? As has already been demonstrated, the challenge to Clausewitz's ideas on the grounds that the state no longer enjoys its place in the structure of international relations is unfounded. Nevertheless, if his Trinitarian formula is not to be thought of as obsolete, it must prove itself against modern examples. In the light of modern challenges to its supremacy, its modern salience will be tested against what are for many the first examples of the 'new wars', the conflicts in Croatia and Bosnia-Herzegovina (1991–95). More importantly, as we have seen, there remain inconsistencies within the formula. As a result, this book questions the idea that the elements within Clausewitz's concept are equal, and re-establishes the centrality of policy – and by extension politics – as the guiding force of war and the principle message of Clausewitz. By assessing the wars in former Yugoslavia by viewing them through a Clausewitzian lens, the book provides a clear indication of the strengths (and weaknesses) of the Trinitarian model and its role in Clausewitzian thought.

Chapter 3
Hostility

'Passion, Hatred and Enmity' and the Wars of Yugoslav Dissolution (1991–95)

The competing components of 'hostility' – passion, hatred and enmity, manifest themselves in myriad ways throughout the course of a particular war. There is no one example which can act as an archetype of hostility. As a consequence of this I do not attempt to uncover each and every instance of passion, hatred and enmity. As noted previously, hostility comprises one of the non-linear elements of war and therefore illuminating its main and recurring features is not a particularly easy task. Apart from anything else, the very nature of non-linearity precludes a neat 'linear' explication. Instead, the purpose of this part of the study is to identify and evaluate the most salient examples of this phenomenon during the wars in former Yugoslavia. How important is hostility? In what way does it influence the war? What inferences can be drawn in regard to its position within the wider analysis of Clausewitz's Trinity? Like the evaluation of chance and uncertainty in the following chapter, hostility can best be understood as part of the non-linear dynamics which make war such a complicated activity. This evaluation of hostility, the purpose of which is to garner a better appreciation of the ways in which it works within the Trinitarian concept, strives to reveal the energy and unpredictability in this component of the triad and highlights both its negative and positive aspects.

Clausewitz tells us that hostile feelings need not necessarily exist prior to the beginning of war. However, his assessment of the Napoleonic Wars suggests that the passions of the people, and especially the escalatory tendencies of these passions, were critically important to any general theory of war. Yet, how did 'hostility' – passion, hatred and enmity – exhibit itself in the Balkan wars? How were the belligerents affected? Clausewitz may have observed that hostile intentions do not require hostile feelings, but even a cursory survey of the process of political fragmentation reveals that hostile feelings had fomented in Yugoslavia in the years and months prior to the breakup of the federal state. The political and social cleavages which Yugoslavia displayed before its disintegration became increasingly deleterious during the late 1980s.[1]

Pre-existing hostility is an important aspect of the deteriorating political situation in Yugoslavia in the years leading to war and can inform the discussion

---

1   For an updated account of Yugoslavia's political fragmentation, see: Dejan Djokic and James Ker-Lindsay (eds), *New Perspectives on Yugoslavia: Key Issues and Controversies* (Abingdon: Routledge, 2010).

of the cause-and-effect relationship of hostility as war is in progress. Despite this, the primary purpose of this chapter is to draw out and debate the salient features of passion, hatred, and enmity as the war evolved. Although the chapter does evaluate existing currents of hostility, it uses this as a stage from which to explore the manifestation of hostility once war is in motion, and subject to the reciprocal interaction which Clausewitz believed so complicated strategy. As such, a more detailed investigation of nationalism in the former Yugoslavia sits outside the purview of this work. Additionally, as the mixture of hostility can vary so enormously from one war to the next, to accurately reveal this element of war's nature is difficult. Hostility can have innumerable consequences and can take equally numerous forms. However, as Clausewitz was clear that war is unique because of actual fighting, the indelible nexus between hostility and violence is a good starting point; the most visual manifestation of hostility is the act of war itself. Although hostility can take different forms in different wars, and even different forms within one particular war, it is appropriate that the study uses Clausewitz as a benchmark on this matter.

As a key characteristic of the Balkan wars is the level of hostility, the problem is not one of locating examples, but of having too many. How does one assess hostility when it seems to envelope everything? To abridge this particular problem the chapter compartmentalizes hostility. Conforming to Clausewitz's own expression, it is divided into two sections, one dealing with hatred and enmity, the other exploring the role of 'passion' and its influence on war's nature. The point here is not to reduce the importance of 'hostility' by reducing into categories. Rather, subdividing these components provides the reader with a sense of how each part of 'hostility' is itself important individually, but also how they can conflate and interweave with each other. Following an essential but brief overview of Yugoslavia's political fragmentation, the chapter explores the influence and impact of hatred and enmity. It then proceeds to explore the meaning of passion. In Clausewitz's comprehension, this element is vital to a more insightful understanding of war's nature, and, as he was acutely aware, it can make the difference between victory and defeat. As a consequence of this, the study draws on the different ways in which passion influenced the nature of the conflicts in Croatia and Bosnia-Herzegovina.

Background to the War: Yugoslavia's Political Fragmentation

The wars in Croatia and Bosnia-Herzegovina (1991–95) were undoubtedly complex, a claim best reflected in the kaleidoscope of political and cultural actors involved in the conflict. It is difficult, at least initially, to differentiate between the belligerents or fully comprehend the motivations for fighting. To clear up some of these problems and to give an insight into the origins of these wars, this section provides a brief introduction and background to Yugoslavia's violent unravelling.

Yugoslavia was first recognized as a state at the Paris Peace Conference in 1919, following the defeat of Germany and her allies at the conclusion of World War One (WWI) in 1918.[2] It lasted as such until the outbreak of World War Two (WWII), collapsing when Yugoslav government forces quickly capitulated to the German high command. In the bloody civil war that followed, the Partisan forces of Tito eventually triumphed, defeating both the royalist Serbian Chetnik, and Croatian Ustashe movements. Supported by the Soviet Union during the war, Tito's Yugoslavia was reborn as a socialist republic after Germany's defeat. Following Tito's split from Stalin in 1948, Yugoslavia embarked on a new course which brought it closer to the West. Leader of the self-styled non-aligned movement, Tito successfully balanced Yugoslavia's delicate National Question and succeeded in positioning Yugoslavia as an important bulwark against Soviet aggression, profiting from Western financial aid in the process.[3]

While the Yugoslav system benefited from this patronage, the state's structure was comprised of a delicate system of checks and balances which aimed at assuaging the strains of the National Question. This worked as long as things were going well, but when the structure was put under strain it was unable to cope. There were several interlinked problems. The end of the Cold War security environment withdrew one of Yugoslavia's unifying roles. Not least, Yugoslavia's economic position was underpinned by US military and economic aid, the International Monetary Fund (IMF) and the World Bank. This money dried up following the end of the Cold War, effectively leaving the state bankrupt. When credit was offered by the international community, it was tied to re-centralization as a means of stabilizing inflation. This was something the republics – particularly Slovenia and Croatia – would not acquiesce to. Further exacerbating these tensions, relatively wealthy republics such as Slovenia and Croatia were taxed to pay for economically weaker republics in the south, causing significant resentment.[4]

---

2   Ivo Banac, *The National Question in Yugoslavia: Origins, History, Politics* (Ithaca: Cornell University Press, 1988), 76–8. For a juxtaposition of the first and second Yugoslav states, see: John Lampe, *Yugoslavia as History: Twice There Was a Country*, revised edition (Cambridge: Cambridge University Press, 2000).

3   The 'National Question' refers to the composition of different nationalities which comprised the Yugoslav state. Each nation held independent political interests and these regularly clashed with the idea of Yugoslav unity. See: Susan Woodward, *Balkan Tragedy: Chaos and Dissolution after the Cold War* (Washington, D.C.: The Brookings Institute, 1995), 21–46.

4   Resentment was compounded by a series of economic crimes perpetrated by those in power. Milošević's theft of over 18 billion dinars may have accelerated the process of fragmentation, and the infamous Agrokomerc affair in Bosnia-Herzegovina, which came to light in 1987, also had political repercussions. Agrokomerc was thought to be an example of solid economic management until it transpired that the board had been issuing false promissory notes – to the cost of $500 million. Fikret Abdić, the then president of the company would resurface in the war in BiH as a major opponent of Izetbegović and leader of the opposition during the 'Muslim' civil war. See: Magaš, Branka *Tracking the Break-*

A key area of republican concern in the 1980s was the future viability of their economies, and by 1990 many in Slovenia and Croatia claimed that Yugoslavia had become the biggest barrier to future economic success.[5] Exacerbating the situation further, by 1989 the rate of inflation had risen to 2,700 per cent per annum.[6] In December 1989, in a move designed to safeguard Yugoslavia's economic legitimacy, Federal Prime Minister Ante Marković implemented a radical anti-inflation policy as a solution to stabilize the economy. Marković's reforms brought inflation down to zero within the allotted six months and the *Dinar*, previously unconvertible, was successfully linked to the Deutschemark. Despite this early triumph, Yugoslavia's industrial output fell by 7 per cent in the first quarter of 1990. For long-term economic stability, Marković needed to redefine the relationship between the republics and the Federal authorities.[7] However, to successfully recentralize the economy, the republics would lose many of the devolved powers they had won in 1974. Moreover, the proposed economic reforms corresponded to Serbian centralist policies. This caused an inevitable impasse, and by the end of December 1990 the federal government's economic system was barely able to function.[8]

The political conflict that led to the fragmentation of Yugoslavia was not about ethnicity and culture per se, although these elements were certainly evident. It was about ideology. More specifically it was about two competing ideas of Yugoslavia, one opting for recentralization; the other for the devolution of powers to the republics. This conflict paralleled national aspirations and was indicative of the ever present National Question. Despite pressure from the centre, a succession of constitutional alterations conceded power to the republics at the expense of the unitary state.

The most famous of these, the 1974 federal constitution, bestowed increased powers on the republics and created two autonomous provinces. Until 1974, Yugoslavia was a federation of six republics: Serbia, Croatia, Slovenia, Bosnia-Herzegovina, Montenegro, and Macedonia. All gained considerably enhanced powers after 1974. Equally importantly, the constitution created the 'new' autonomous provinces of Vojvodina and Kosovo – both provinces were constituent

---

Up 1980–92 (London: Verso, 1993), 111–12. These anxieties were compounded again following Milošević's theft of $1.8 billon to shore up Serbia's crumbling financial system. *Tanjug*, 8th January 1991, SWB EE/0967 II.

    5   Z. Lazarević, 'Economic History of Twentieth Century Slovenia', J. Benderly and E. Kraft (eds), *Independent Slovenia* (New York: St Martins Press, 1994), 69–90; James Gow and Cathy Carmichael, *Slovenia and the Slovenes* (London: Hurst & Co., 2000), 112.

    6   Economist Intelligence Unit, 1989; *Financial Times*, 6 July 1990, 35.

    7   Woodward, *Balkan Tragedy*, 80–81, 100.

    8   By as early as 1988, opinion polls showed that around 60 per cent of Slovenia's population thought it could develop faster if it dissolved its relationship with Yugoslavia. The situation was mirrored in Croatia. See: *Tanjug*, 16th February 1988, SWB EE/0082 B/11.

parts of the Republic of Serbia.[9] Though termed 'autonomous provinces', in reality both held similar powers to the republics. Although the new constitution was intended as a solution to the 'National Question', retrospectively it is clear that it created deeper fissures throughout the unitary state, which eventually resulted in its dissolution.

When Tito died in 1980, the rotating presidency, which was intended to allay republican apprehensions, failed and the state was gradually gripped by rising nationalism.[10] Although these anxieties had never disappeared, this new wave of tension was instigated by the Serbs. This related directly to the 1974 constitution and the loss of Serb authority over Kosovo – the locus of Serbian culture and identity. In 1986, the now notorious Serbian Academy of Arts and Sciences (SANU) memorandum was leaked to the press. The memorandum stressed the increasingly dire state of the Serbs in Kosovo – Albanian intimidation and ethnic intolerance of Serbs – and condemned the 1974 constitution for dividing the Serb nation. The Yugoslav body politic was rocked further by the coup-style takeover of the League of Communists of Serbia (LCS) by Slobodan Milošević in 1987. Following his infamous proclamation to the Serbs of Kosovo that 'nobody should dare to beat you' his rise to prominence was rapid, usurping power from his mentor and one-time friend Ivan Stambolić.[11]

This was followed by a political programme designed to recentralize the Serb state. Orchestrated by Milošević, this 'anti-bureaucratic revolution' provoked the annexation of Kosovo and Vojvodina in 1988. Montenegro followed in 1989, and the Serbian Assembly declared a new constitution on 28th March 1989. Growing republican suspicion that Milošević was creating a 'Serbo-Slavia' became even more plausible with the acquisition of these additional Federal votes and put Slovenia and Croatia on a collision course with Serbia.[12] Milošević was now in a position to overrule the Federal Presidency to gain Serbia's political aims: Yugoslavia's Federal structures had irreversibly changed.

Slovenia recognized that it would be unable to proceed with its own political reform within Yugoslavia and by September 1989 it had projected a list of

9   'The Constitution of The Socialist Federal Republic of Yugoslavia, Promulgated on 21st February 1974', in Snezana Trifunovška (ed.), *Yugoslavia Through Documents From Its Creation to its Dissolution* (Dordecht: Martinus Nijhoff Publishers, 1994), 224–33.

10   The Presidency finally became defunct when Serbia refused to acknowledge the Presidency of Stipe Mesić. An interesting examination of the unfolding drama comes from Mesić himself: Stipe Mesić, *Demise of Yugoslavia: A Political Memoir* (Budapest: Central European University Press, 2004).

11   Slavoljub Dukic and Alex Dubinsky, *Milošević and Marković: The End of the Serbian Fairytale* (Montreal: McGill University Press, 2001), 22–33.

12   Transcript of the interview, Milan Kučan, Slovene Secretary of the Central Committee of the Communist Party 1986–90, and first President of Slovena, 1990. Conducted for the Brook Lapping television series *The Death of Yugoslavia* (first broadcast on the BBC September 1995), 3/45. Transcript deposited in Liddell Hart Centre for Military Archives (hereafter – LHCMA).

amendments to the 1974 constitution – including the right to secede and the right to allocate the republic's wealth to the benefit of Slovenia.[13] Though experiencing significant federal pressure, Slovenia went ahead with its planned alterations following Croatian support at a meeting of the Central Committee of the Yugoslav Party. Slovenia declared itself a sovereign state on 27th September 1989. By the time Slovenia and Croatia walked out of the now infamous 'Fourteenth Extraordinary Party Congress' on 23rd January 1990, the damage had already been done.[14] After democratic elections in Slovenia and Croatia, both republics voted for secession in national plebiscites. They were officially declared independent on 25 June 1991.[15] On the 26 June, the Federal government proclaimed the declarations illegal, ordering the Yugoslav army to seize the borders crossings and other strategic centres in an attempt to halt secession; triggering violent conflict in the process. However, after ten days of fighting the Slovenes secured their prize – independence.

A more complicated situation had been building in Croatia. Twelve per cent of the Croat population was Serb, living mostly in the Krajina region – the old military border of the Austro-Hungarian Empire. As the strains in the federal structures of the state began to turn into more visible political fissures, the Serbian population in Croatia looked to Serbia for support and protection; Serbian policy was to actively stoke resentment as a prelude to annexation. Although the Yugoslav National Army (JNA) maintained a veneer of impartiality, it actively armed Serb militias in preparation for combat.[16] In an act that would have repercussions throughout the coming war, during May 1990 the JNA issued instructions to the republics that all Territorial Defence (TO) weapons must be handed over to Yugoslav military authorities – Serb populations were excused the ultimatum.[17]

---

13   Silber and Little, *Death of Yugoslavia*, 74; Gow and Carmichael, *Slovenia and the Slovenes*, 167–74.

14   At the conference Slovenia advanced the unlikely outcome of remaining in an asymmetric federation, increasing its autonomy within Yugoslavia but remaining within the territorial boundaries of the Federation. Following the walkout, the Slovene leadership disbanded the League of Communists of Slovenia. By undertaking this action they also effectively dissolved the LCY and, therefore, the very structure that united Yugoslavia as a unitary state. The only credible Yugoslav organization remaining was the JNA. For a recent investigation of Slovene motives at the conference, see Dejan Jović, 'The Slovenian-Croatian Confederal Proposal: A Tactical Move or an Ultimate Solution?', in: Lenard J. Cohen and Jasna Dragović-Soso, *State Collapse in South-Eastern Europe: New Perspectives on Yugoslavia's Disintegration* (West Lafayette, Indiana: Purdue University Press, 2008), 249–80.

15   *Tanjug*, 2nd July 1990, SWB EE/0807 B/8.

16   By 1988, the military had decided that Serbia offered the best chance of preserving and enhancing Yugoslav centralism. James Gow, *Legitimacy and the Military: The Yugoslav Crisis* (London: Palgrave Macmillan, 1992), 56.

17   Laura Silber and Alan Little, *The Death of Yugoslavia* (London: Penguin, 1995), 105–7.

This highly successful policy guaranteed that Serb irregulars and the Serb army would enjoy a quantitative and material superiority in weaponry throughout the conflict, shaping the character of the war before the real fighting had even begun.

Although Slovenia was ethnically homogenous, Croatia and Bosnia-Herzegovina had large ethnic minorities. President Tudjman's Croatian Democratic Community (HDZ) won Croatia's first free elections in April 1990, and had come to power on the back of his promise to make Croatia a state of the Croatian nation.[18] After his inauguration as President he began the process of writing a new constitution. The wholly Croatian character of the constitution served as a catalyst for the more radical Croatian Serbs, paving the way for wartime leader, Milan Babić, who had been co-opted by Belgrade, to 'defend Serbian independence'. On 1st April 1991 the executive council of the region issued a statement proclaiming that 'the Serbian autonomous region of the Krajina is joining the Republic of Serbia'.[19] By August that year the conflict was escalating and becoming increasingly brutal. As the Vance Plan began the deployment of a UN peace force (UNPROFOR) to Croatia, ostensibly stabilizing the situation, the war moved into Bosnia-Herzegovina (BiH) in early 1992. The political composition of BiH following elections in the republic in December 1990 resulted in a breakdown of 99 Muslim seats, 85 Serb, and 49 Croat, reflecting the ethnic demographic of the republic.[20] Izetbegović led a government of 'national unity', which incorporated the three main ethnic communities into government, but this was put under mounting strain by Serb and Croat agitation – by May 1991, the Bosnian SDS started to demand secession so as to link with Croatia's Serbs, making cohesive government impossible. The combination of internal unrest and external agitation deepened existing fractures. When the results of Bosnia's referendum on independence were published on 2nd March the following year, the Bosnian government, led still by Izetbegović, was in an impossible position, locked between two increasingly likely adversaries. War followed quickly.[21]

---

18   Tudjman secured 41.3 per cent of the first vote, 42.2 per cent in a second round, securing 54 seats in the Sabor (67.5 per cent of the seats). The Serbian SDS largely boycotted the election and secured just one seat – although 46,418 did vote. Cited in Lenard J. Cohen, *Broken Bonds, Yugoslavia's Disintegration and Balkan Politics in Transition*, 2nd edition (Boulder: Westview Press, 1995), 100.

19   *The Times*, 2 April 1991, 7.

20   Noel Malcolm, *A Short History of Bosnia*, new updated edition (New York: New York University Press, 1996), 222.

21   Although the Serb and Croat sections of society did not vote, over 92 per cent of those voting did so for independence. On Bosnia, Lenard J. Cohen, *Serpent in the Bosom: The Rise and Fall of Slobodan Milošević* (Boulder: Westview Press, 2002) provides a thorough analysis of the political and security currents. For surveys of the political and cultural currents in Bosnia-Herzegovina, see: Malcolm, *A Short History of Bosnia*; and Robert J. Donia and John V.A. Fine Jr. *Bosnia-Herzegovina: A Tradition Betrayed* (New York: Columbia University Press, 1994).

The political fragmentation of Yugoslavia and the subsequent wars of secession resulted from complex interrelated causes. They were not the result of a spontaneous explosion of ethnic hatred, although ethnic dissonance certainly has a role in the story. Equally importantly, neither was the war solely about elite predation, as argued by many of the new war advocates. Although elites tapped into cultural veins to promote their competing visions of the future, ethnic tension did not arrive suddenly at the whim of leaders such as Milošević.[22] The National Question had never disappeared completely from Yugoslav politics, and the combination of decentralization, loss of unifying ties with the end of the Cold War, and more particularly the collapse of the economy, all contributed to the eventual ruin of the second Yugoslav state. Interestingly, if not a little ironically, the violent dissolution appears to have been partly a response to the complicated layering of sovereignty. This complex underwriting of citizen rights had created a web of layered loyalties. This worked well to alleviate suspicion by providing rights to citizens during the Titoist era. However, if the state was to dissolve, it was far from clear what group actually held sovereignty.[23] With so much power vested in the republics after 1974, the republican borders began to be viewed as inviolable – as they were by the international community. The resulting war was effectively about who owned what; as both Croatia and Bosnia had large ethnic minorities which agitated for secession this was a serious problem.[24]

As noted already, the acute feelings of animus displayed by the belligerents and the extent of war crimes gripped global media attention and drew the international community into the crisis. In a world stripped of Cold War anxieties it was also a crisis that the UN could conceivably manage without the overt threat of escalation into nuclear war. Despite being involved in one shape or form from the initial phases of the war in Slovenia, the external powers were beset by their own interests and never fully understood the motivations of the belligerents; for outside observers the war seemed to be about ethnic hatreds rather than political interest. This resulted in a catalogue of errors by the international community, which only averted complete failure after renewed US interest in the Clinton era. All of these factors played into the war, producing a strategic environment where it was profitable to gain maximum concessions from one's opponent.

---

22   This point is also made by Stathis Kalyvas and Nicholas Sambanis. See: Stathis N. Kalyvas, and Nicholas Sambanis, 'Bosnia's Civil War: Origins and Violence Dynamics', in Paul Collier and Nicholas Sambanis (eds), *Understanding Civil War: Evidence and Analysis* (Washington, D.C.: The World Bank, 2005), Vol. 2, 191–229.

23   Woodward, *Balkan Tragedy*, 41–4.

24   Bosnian Census 1991 'Federation of Bosnia-I-Herzegovina Federal Office of Statistics'. In 1991, census figures demonstrate the republic's demographic make-up of the three principle groups as follows: Muslims 43.5 per cent, Serbs 31.2 per cent, and Croats 17.4 per cent. Although in 1991, the Muslim nation formed a majority; the 1961 census demonstrates that Serbs were the majority nation, holding 42.9 per cent of the population compared to the Muslims' 25.7 per cent and the Croats' 21.7.

In BiH in particular, the three-way split between Serbs, Croats, and Muslims, and their occasional willingness to realign their allegiances, was seen as evidence of duplicity.

Hatred and Enmity

The Rise of Hostility in Yugoslavia

There is little doubt that acute levels of hostility marked a key characteristic of the wars of former Yugoslavia, making them identifiable through their brutality and cultural intolerance. Although hostility was generated by the war, it was underpinned by existing animus which percolated through the Yugoslav system in the years and months before dissolution. The rising tide of nationalism in the 1980s was a harbinger of the hostility that would engulf Yugoslavia as its dissolution became violent. As noted above, this aspect of the Yugoslav crisis is important not simply because it casts light onto why Yugoslavia disintegrated in the way that it did; existing hostility also provides the context in which the war took place, underpinning the ethnic/cultural narrative.

A key component of the SANU memorandum was the imagined genocide of the Serbian people in Kosovo and the requirement for Serbia to reorganize the constitution. In articulating this message, Serbian nationalists tapped into pre-existing Serbian notions of Serbian nationalist historiography and culture, purposefully re-awakening the anxieties which accompanied them. The Serbs' identification as a national group has been expressed through the Kosovo myths and the 'Serbian Epics', which tell the story of Serbia's demise as a medieval state; the most famous narrates the story of the Serbian defeat by the Ottomans at the battle of Kosovo on 29th June 1389. In 'The Downfall of the Kingdom of Serbia' – Prince Lazar, the leader of the Serbian army, was met in a dream by St Elias, who offered him a choice between the 'Earthly' or the 'Heavenly' Kingdom. Prince Lazar chose death and eternal salvation by opting for the 'Heavenly Kingdom'. David Macdonald explains that the choice represents 'the victory of the divine over the secular, the eternal over the temporal … Like the crucifixion, the martyrdom of Lazar and the Serbian nation raised the Serbian people and made them divine, holy, chosen, special.'[25] This symbolism is intimately entwined in Serbian cultural historiography. As Vjekoslav Perica observes, 'Kosovo is the central myth and the symbol of Serbian Orthodoxy. It is also a sacred centre, the nation's rallying point

---

25   David Bruce Macdonald, *Balkan Holocausts?: Serbian and Croatian Victim Centred Propaganda and the war in Yugoslavia* (Manchester: Manchester University Press, 2003), 70. For more information on this aspect of Serbian historiography, see: Branimir Anzulovic, *Heavenly Serbia: From Myth to Genocide* (London: Hurst & Co., 1999); and Tim Judah, *The Serbs, History, Myth and the Destruction of Yugoslavia*, second edition (New Haven: Yale University Press, 2000).

and its lost paradise.'[26] When the Kosovo issue and the wider 'Serbian Question' (this refers to the position of Kosovo as an autonomous province) was revived during the 1980s, the myths associated with Kosovo were reinvented and the old story was provided with a modern context.

Although the rise in support for Serbian nationalism pre-dated the Milošević phenomenon, there is consensus that he was the first to embrace nationalism as a political ideology as a means of winning power in a period of transition. In the words of Warren Zimmerman, the last US ambassador to Yugoslavia, he was 'an opportunist rather than an ideologue, a man driven by power rather than nationalism'.[27] Milošević's rise to political prominence coincided with the downturn in Yugoslavia's fortunes and he is widely viewed as one of the first to recognize the changing political landscape in Yugoslav politics. The beginning of this transformation is intimately linked to Kosovo. Sent to the province as a subordinate of Ivan Stambolić, the then President of Serbia, Milošević profited from the depth of inter-community suspicion and anxiety evident in the province during the 1980s.[28] Addressing a gathered crowd of Serbs protesting against 'institutional' Albanian persecution, Milošević called out to them that 'no one shall dare to beat you'.[29] He became a Serbian hero overnight and used his new image to reinvent himself as defender of the Serbian nation, usurping power from his former mentor Stambolić in the coup-style takeover of the League of Communists of Serbia (LCS) in 1987.

During 1988 and 1989 Milošević's 'anti-bureaucratic revolution', or 'happenings of the people' as they were also known, further cemented his power.[30]

---

26   Vjekoslav Perica, *Balkan Idols, Religion and Nationalism in Yugoslav States* (Oxford: Oxford University Press, 2002), 8.

27   Warren Zimmerman, *Origins of a Catastrophe* (New York: Random House, 1999), 25.

28   Although the 1974 Constitution made Kosovo an 'autonomous province' within Serbia, unrest and major rioting took place as early as 1981, polarizing still further the two major ethnic communities: Serbs and Albanians. See: Noel Malcolm, *Kosovo, A Short History* (London: Macmillan, 1998), 334–56. For more on Milošević's role in Kosovo, see: Tim Judah, *Kosovo, War and Revenge* (New Haven: Yale University Press, 2002), especially pages 33–60.

29   Although the demonstration appeared to be spontaneous, in fact it had been pre-arranged by Milošević supporters. Robert Thomas, *The Politics of Serbia in the 1990s* (New York: Hurst & Co., 1999), 44.

30   For a recent assessment of this period, see: Nebojša Vladisavljević, *Serbia's Anti-bureaucratic Revolution: Milošević, the Fall of Communism and National Mobilization* (Basingstoke; Palgrave Macmillan, 2008). For an updated account of Yugoslavia's political fragmentation, see: Nebojša Vladisavljević, 'The Break-Up of Yugoslavia: The Role of Popular Politics', in Dejan Djokic and James Ker-Lindsay (eds), *New Perspectives on Yugoslavia: Key Issues and Controversies*; and Ivo Banac, 'The Politics of National Homogeneity', in Brad K. Blitz (ed.), *War and Change in the Balkans: Nationalism, Conflict and Cooperation* (Cambridge: Cambridge University Press, 2006), 30–43.

By tapping into the Kosovo legend he was able to pull support from a wide constituency, reclaiming the political powers devolved to Kosovo and Vojvodina in the 1974 constitutional alterations. In 1989, pro-Milošević supporters gained control of the Montenegrin Presidency; effectively providing Milošević with de facto control of that republic. It is important that the use of emotive symbolism and the general rise in hostility is assessed in conjunction with the wider political and economic problems facing Yugoslavia at the end of the 1980s. The stage was set for a revival of Serbian nationalism and Milošević skilfully exploited existing currents of opinion, especially the idea that the Serbs were 'again' under threat. The tacit support offered to Milošević's growing national programme by the Serb Church, as well as the intellectual elite, provided the regime with the credibility it needed.

The historic union between church and state, and the support invested in this relationship by nationalist intellectuals, Dobrica Ćosić in particular, was an important factor when the focus of political events moved from Kosovo to Croatia and Bosnia-Herzegovina.[31] When attention did turn to Croatia and Bosnia-Herzegovina, nationalist sentiments were propounded on a daily basis and the official line was underpinned by its association with Serbia's cultural elite and religious institutions. Coupled with his dominance of the media, Milošević had free rein with which to propound Serbia's new nationalist rhetoric – manipulating Serbia's historical narrative as a means of underpinning power was an important feature in the rise of hostility.[32]

In one of the most obvious examples of the modern manipulation of the Serbian narrative, Milošević addressed those attending the 600th anniversary of the battle of Kosovo on 28 June 1989. Evoking the memory and spirit of the Kosovo legend to a Serbian crowd of over one million, he sent a message of his intentions to his opponents in the other Yugoslav republics. The celebration at Gazimestan, the site of Serbia's medieval defeat by the Turks, was the culmination of a politically inspired Serbian cultural awakening, which in the weeks before had seen the bones of Prince Lazar toured throughout the Yugoslavia republics.[33] Although the celebration marked the 600th anniversary of the battle of Kosovo, the message was directed at those republics which Milošević suspected of conspiring to secede

31   Ćosić was a leading figure in the nationalist movement and a respected novelist and intellectual; he became president of the Federal Republic of Yugoslavia in 1992. The church/state relationship predated the battle of Kosovo, with the church enjoying the position of a 'quasi-political institution'. Anzulovic, *Heavenly Serbia*, 22.

32   Eric Gordy has called this the 'destruction of alternatives', a government policy aimed at maximizing and regaining control through stemming information and non-government outlets of expression. See: Eric Gordy, *The Culture of Power in Serbia: Nationalism and the Destructions of Alternatives* (Pennsylvania: Pennsylvania State University Press, 1999), 61–10.

33   The Slovene authorities banned a rally planned for Slovenia. In a poll carried out by the daily Slovene newspaper *Delo*, 80 per cent were against the rally taking place in Slovenia. *Belgarde Home Service*, 5 March 1989, SWB EE/0402 B/10.

from Yugoslavia – in the process separating large Serbian minorities from Serbia. The St Vitus day celebrations invoked a visceral cultural memory, acting as a portent of what was to come.[34]

When the political limelight moved to the Croatian position on its secession from Yugoslavia, it was easy for Milošević to present the position of the Krajina Serbs in the starkest form: not only were the Serbs losing their status as equal citizens, they were in danger of being driven from their lands. Although the Serbian government had serious concerns regarding the political situation of the Serb community in Croatia, Serbian propaganda propounded the line that Serbs were in danger from a new wave of Croatian revanchism. With the Ustashe terror still within living memory, it was easy to fuse the Kosovo legend with a modern story.[35]

Led initially by Jovan Rašković, Croatia's Serbs had been worried by the increasingly polarized views expressed by Serbia, Slovenia and Croatia. Although Milošević had signalled that the Slovenes could secede if they so wished, with such a large Serb population in Croatia, Tudjman's government faced much stiffer opposition. If Croatia seceded from Yugoslavia the Serbs would become a minority in a state which had previously committed genocide against it. There was a tangible sense of unease that Croatian independence would lead to political and cultural repression of the Serb community. As the ideological crisis threatened to split the state, and as Croatia proceeded with its political reform, the dynamics of the security situation became more aggressive.

Croatia's national revival was a response to the new wave of Serbian nationalism and irredentism which had swept the republic – and Yugoslavia – in the late 1980s. This new sense of Croatian nationalism engendered considerable apprehension among its Serb minority, who feared a return to the persecution of the wartime Ustashe government. As republican antipathy grew, so did the move towards identity politics.[36] Croatia openly expressed its desire to be an

---

34   David Macdonald has identified the use of 'Fall' and 'Revival' as key to the inculcation of the Serb historical narrative and its manipulation during the years of Yugoslav political dissolution. David MacDonald, *Balkan Holocausts?*, 63–97. For a recent examination of the role of identity, see: Pavlos Hatzopoulos, *The Balkans beyond Nationalism and Identity, International Relations and Ideology* (London: I.B. Tauris, 2008).

35   Macdonald, Ibid., 80.

36   The articulation of ultra-nationalist Serbian politics should not be underestimated as a significant motor to the rise of Croatian nationalism. According to myths associated with Serbian ultra-nationalism, there are varying explanations of the Serbian nation. Made popular before and during the conflict was the notion that 'where our forebears' bones now rest, there the Serbian frontiers stretch'. In other words, the Serbian nation exists wherever there are Serbs graves. This evoked a particularly potent message and represented the political fallout vis-à-vis the genocide of Serbs at the hands of the Croatian *Ustashe* during WWII. Other myths centre round the idea of Serbia as 'mother', and the lands inhabited outside Serbia proper as her children. In yet another depiction, Serbia is imagined as a human body. The soul lies in Kosovo, the brain in Belgrade. This was a particularly emotive myth prior to the outbreak of war. The large Serbian territories outside Serbia were seen as

independent state. The very success of this revival, and of Tudjman's new political party, rested on the assumption that Croatia would become a nation-state. It refused to recognize that Serbs living in the Krajina had genuine fears about their cultural future, especially the prospect that their ties with Serbia would be severed. Croats, who had only recently found their national voice, reverted to Croatian symbols to express their national consciousness. Many of these symbols had hitherto been banned. The *Šahovnica* in particular – Croatia's checkered flag, though a genuine emblem of the medieval Croatian state, was also the flag of the Ustashe. For Serbs living in Croatia, its symbolism was a potent reminder of earlier periods of national anomie.

Stories about mass graves began to permeate the local and international media and would play a major part in selling the war as ethnically inspired. Indeed, Serbia would come to rely, even once war had begun, on Croatia's wartime past. In propaganda terms, the Serbian leadership would relate this back to the Jasenovac concentration camp, distilling a feeling of national sacrifice akin to the 'Field of Blackbirds' in Kosovo.[37] Of course, to the outside world these tactics would serve only to confirm Yugoslavia's bloody history had conspired against it once again. Furthermore, when groups assumed the trappings of cultural nationalism as a means of displaying their identity, this only served to polarize the political agenda even further.

In Bosnia, the situation was potentially even more explosive and although it had managed to partly isolate itself from the escalating tension between Serbia, Slovenia and Croatia, its ethnic composition meant that it could not escape the violence. In terms of hatred and enmity, the Bosnian case is intriguing. During 1990, opinion polls commissioned by the federal government suggested a rejection of the various nationalist programmes, but the success of nationalist parties in the other republics made the situation in BiH incredibly tense, provoking people to vote according to ethnicity.[38] When elections did take place communities voted along ethnic lines, polarizing the republic as a result. Although Izetbegović had been imprisoned for alleged Islamic fundamentalism following the publication of his treatise the *Islamic Declaration*, Islamic practice was relaxed and only became more militant as a response to the conflict. Izetbegović tried to assuage tensions by leading a multi-party government, but the republic was already gripped by the

---

the arteries of the Nation, and losing them reflected amputation. Interestingly, during the schism between the Bosnian Serbs and Belgrade, Karadžić reflected on the fluid nature of the Nation, that is, its ability to reconfigure. Following the schism, the centre of the Serbian nation is not in Kosovo or Belgrade, but Republika Srpska. See: Ivan Čolović, *The Politics of Symbol in Serbia* (London: Hurst & Co., 2002), especially, 9, and 29–38.

37   Estimates of the numbers of Serbs killed at Jasenovac vary according to national group and range from 80,000 to over one million dead. This became a major cause of disquiet between Serbia and Croatia in the run up to hostilities.

38   Cohen, *Broken Bonds*, 103.

emerging anarchy of civil war. With the loss of social norms, violence erupted.[39] Although Bosnia tried to prevent the escalating hostility, outside pressures ensured that averting war was impossible. Like Croatia, when war did break out, the convergence of politics, opportunism, and history collided to produce an inflammable mix of hostility.

The Reciprocity of Hatred and Enmity

Existing hostility was exacerbated by the reciprocal effects of conflict. Although groups had already been retreating into the semi-safety of ethnic and cultural communities before the conflicts began, the outbreak of hostilities dramatically increased this process. During the period 1991–95 a total of 1,024 mosques – as well as other sites of cultural and religious significance: 182 Catholic churches, and at least 28 Serb Orthodox places of worship – were destroyed.[40] In terms of the Bosnian conflict alone, although the real figure sits at approximately 100,000, the initial numbers of dead were estimated to be as high as 250,000 and figures for population displacement were many times this number.[41] The war was destructive not just in the sense that many people were killed, but also because the levels of hostility and the violence inflicted on other ethnic groups seem so visceral. Although the perception of the war as purely ethnically driven and therefore irrational is a false one, it is not hard to see why these conflicts are frequently assumed to represent the epitome of religious and cultural hatred.[42] Attacks against religious symbols were an attack at the heart of the opponent's

---

39   Barry Posen, 'The Security Dilemma and Ethnic Conflict', *Survival*, 35(1) (Spring 1993), 27; see also: Jack Snyder and Robert Jervis, 'Civil War and the Security Dilemma', in B.F. Walter and J. Snyder, (eds), *Civil Wars, Insecurity, And Intervention* (New York: Columbia University Press, 1999), 15–37. Other important works pertaining to ethnic war include: Michael Brown, 'Causes and Implications of Ethnic Conflict', in Michael E. Brown (ed.), *Ethnic Conflict and International Security* (Princeton: Princeton University Press, 1993); David Welsh, 'Domestic Politics and Ethnic Conflict', *Survival*, 35(1) (Spring 1993), 63–90; David A. Lake and Donald Rothchild, 'Containing Fear: The Origins and Management of Ethnic Conflict', *International Security*, 21(2) (Fall 1996), 41–75.

40   Perica, *Balkan Idols*, 166.

41   Figures for the number of dead are disputed, however, the International Criminal Tribunal for the Former Yugoslavia estimates that 102,622 died. This roughly mirrors the Sarajevo Research and Documentation Centre (SDC) estimation of 97,207, half the original estimate. Details are available at: http://www.idc.org.ba/presentation/index.htm (accessed 22 July 2008).

42   Interestingly, although the new war idea is constructed around a key claim that war has become de-institutionalized and where the targets of choice are civilian, the breakdown of casualties provided by the SDC confirms that the number of civilians who lost their life was, although high, still significantly lower than that of the number of soldiers killed. The number of civilian dead during the period 1991–95 was 39,685 (21,814 of whom died in 1991). Casualties for soldiers over the same period were 57,523 (23,297 of whom died in 1991).

identity; their purpose was to demoralize and ethnically cleanse territory. Identity was viewed as a 'centre of gravity', which if destroyed could herald the end of the enemy's will to resist – hardly a new tactic.[43] Sabrina Ramet notes,

> [The] Serbian Insurrectionary War was not about religion but about hegemony. When Serbs blew up mosques and Catholic churches and when Croats destroyed mosques and other religious buildings, they were not, in fact, doing so to spread their own faith, but rather to destroy the artifacts which established other peoples' history in the area and which helped members of other nationalities remember their past and hold on to their cultural identity. In other words, attacks on religious objects served political purposes; politics was primary, not religion.[44]

In Croatia and Bosnia, the Serbian strategy was to annex Serbian areas for incorporation into a remodelled Yugoslavia. The initial tactic was to use irregular troops and paramilitary organizations. Groups such as Arkan's 'Tigers', Vojislav Šešelj's Chetnik organization, Berli Orlovi – the White Eagles – and the Knindjas were used in tandem with village militias to pave the way for the annexation of Serb territory. These groups intimidated the local non-Serb population *in terrorem* – through murder, beatings and rape.[45] They also recruited and trained local volunteers, further severing local ties and friendships.[46] As the violence increased,

---

See: http://www.idc.org.ba/presentation/index.htm (accessed 19 June 2008), research results for 'civilians' and 'soldier'.

43   The Clausewitzian concept of the 'Centre of Gravity' (CoG) has been described as the bridge between politics and strategy. The CoG can be described as the hub of the enemy's power; this could be a capital city, a stronger ally, or its will to resist. Attacking it successfully is seen as a way of decisively defeating the enemy. See: Carl von Clausewitz (1882), *On War*, transl. by Michael Howard and Peter Paret (New York: Alfred A. Knopf, Everyman's Library edition, 1993), 746–9.

44   Sabrina P. Ramet, *Balkan Babel, The Disintegration of Yugoslavia From The Death of Tito to the Fall of Milošević*, Fourth Edition (Boulder: Westview Press, 2002), 81; Silber and Little, *The Death of Yugoslavia*, 83. Although the war magnified cultural links, these already existed. In Croatia the suppression of the Croatian Spring meant that the only way Croats could display nationalist sentiment was through their relationship with the church. Perica, *Balkan Idols*, 59. Interestingly the close relationship with the church in Croatia mirrors the importance attached to the church in Serbia. During the years of Ottoman control, the church was the last bastion of Serbian nationhood. The church performed a vital role in both cases. See: Anzulovic, *Heavenly Serbia*, 23–31.

45   See Kalyvas, *The Origins of Violence in Civil Wars.*

46   These groups have become one of the symbols of the conflict and were thought to validate the notion that the war was disorganized. It has since transpired that many of these groups were funded and trained by government agencies – of the five Serb groups active during the war, four were financed by the Serb secret police. During his interview for the Death of Yugoslavia television documentary Šešelj exposes the Yugoslav and later Serbian government's provision of training camps, military equipment and money, and discusses joint operations between paramilitary organizations and the regular army.

the ethnic displacement of the Croat and Bosnian populations intensified.[47] So too did the cycle of hostility and hatred, with hatred manifesting itself in multiple ways, provoked by the reciprocity of the war.

Undoubtedly, people killed for a variety of reasons and national pride did not always top the list. For some, the war provided the chance to exact revenge for past wrongs, many of which were personal rather than ethnic. For example, as Kemel Pervanic notes, many of the deaths of inmates in prison camps were the result of a grudge rather than overarching national programmes.[48] In fact, as John Mueller has demonstrated, much of the killing throughout the war can be attributed to a relatively small number of people.[49] As demonstrated in more detail below, inspiring soldiers to fight is not always easy, and in the Serbian case the worst jobs were kept for the paramilitary organizations, whose cadres were composed of disillusioned young men, criminals freed from prison, and football hooligans indoctrinated into ultra-nationalist factions by figures such as Arkan.[50] Nevertheless, it is axiomatic that reciprocal animosity played a crucial role in permeating and undermining old loyalties, contributing to the societal breakdown which characterized the conflict in Bosnia. There are plenty of examples of volunteers joining to revenge the deaths of family members killed in the war.[51] This type of reciprocity is also highlighted in studies of rape during the conflict. In one study of war crimes perpetrated against women, Kelly Askin writes that:

> Revenge is multifaceted and notoriously circular, weaving layers of hatred and destruction. Notably, Croats and Muslims are raping women in retaliation for Serb rapes upon non-Serb women, and more Serbs are raping in retaliation for the rape of Serb women. Men revenge the rape of 'their' women by raping the enemy's women, whose men revenge their rape by raping even more women, whose men revenge their rape by raping still more women even more viciously.[52]

The long-term effects of brutality such as this significantly undermined any prospect of quick rapprochement. In another high-profile example, one of the

---

Transcript of the Interview, Vojislav Šešelj, 'Leader of Serbian Chetnik Paramilitary Organisation'. Conducted for the Brook Lapping television series *The Death of Yugoslavia*, 3/69. Transcript deposited in the LHCMA.

47   'Death Toll Climbs in Eastern Croatia as Yugoslav Troops, Militiamen Clash', *The Washington Post*, August 26, 1991.

48   Kemal Pervanic, *The Killing Days* (London: Blake, 1999), 156–7.

49   John Mueller, 'The Banality of Ethnic War', *International Security*, 25(1) (Summer, 2000), 42–70.

50   See: Čolović, *The Politics of Symbol in Serbia*, 259–87.

51   Roger Cohen, *Hearts Grown Brutal: Sagas of Sarajevo* (New York: Random House, 1998), 137; Chuck Sudetic, *Blood and Vengeance: One Family's Story of the War in Bosnia* (New York: Norton, 1998), 157.

52   Kelly Dawn Askin, *War Crimes against Women: Prosecution in International War Crimes Tribunals* (The Hague: Martinus Nighoff Publishers, 1997), 287.

reasons posited for the murder of 7,000 men and boys at Srebrenica in July 1995 was revenge.[53] The town had come perilously close to destruction in 1993, surviving only because it had attained safe-haven status at the very last moment. It was not so lucky in 1995. The Serb attack began on 6 July and the battle was over by 11 July. By 13 July over 7,000 men and children had been murdered in the biggest single act of genocide since World War Two. With the military momentum now seemingly with the Serbs, they stepped up their offensive with the assault of another 'safe-area', Žepa.[54] How can this level of violence be explained?

The violence meted out at Srebrenica seems to have been motivated by interlinked causes. Although the town was a valid strategic target, the rationale behind the slaughter remains unclear. At the time of the attack in July 1995, the Bosnian-Serb Army (VRS) was starting to meet stiffened resistance from the increasingly cohesive partnership between the Croatian Defence Council (HVO) and ARBiH, and Srebrenica may have been partially a reaction to this. Additionally, though NATO was slowly becoming involved in the crisis, the Serbs appear to have disregarded the seriousness of NATO ultimatums; a notion no doubt emboldened by NATO's poor showing during this crisis. What did the VRS get out of it? Clearly they could have won the enclave without the murder of so many innocent lives. A heady mix of revenge and intransigence appears to be the most significant motive for the massacre.[55] One of three Muslim enclaves within Republika Srpska, it had been a thorn in the side of Serbian forces from the start of the war and the Muslim defenders of the town frequently raided outlying Serbian villages such as Podravnje, Grabovacka, and Kravica. This was despite the fact that the official classification of Srebrenica as a United Nations (UN) 'safe-haven' prohibited military operations from the enclave. These raids were brutal and appear to have been contributing factors in the causal chain which led to the genocide. Interestingly, just as the Serbs appear to have been reacting to

53   'Bosnia-Herzegovina – The Fall of Srebrenica and the Failing of UN Peacekeeping', *Human Rights Watch*, 7(13), October 1995.   Naser Orić, the Muslim commander of Srebrenica was indicted for his part in crimes against humanity perpetrated against the Serb population around Srebrenica. See: The International Criminal Tribunal for the Former Yugoslavia. *The Prosecutor vs. Naser Orić* – Case No. IT-03-68-T.

54   See: Jan Willem Honig and Norbert Both, *Srebrenica, Record of a War Crime* (London: Penguin Books, 1996).

55   In one revealing incident, Owen explains that Milošević's warned that 'if the Bosnian Serb troops entered Srebrenica there would be a bloodbath because of the tremendous bad blood that existed between the armies.' Naser Orić was the ARBiH commander responsible for a massacre near Bratunac in 1992.' David Owen, *Balkan Odyssey* (New York: Harcourt Brace & Company, 1995), 135. Nevertheless, it is important to note that the killings were not spontaneous. UN investigations clearly demonstrate the careful planning that went into the killings. See: United Nations, 'Report of the Secretary-General pursuant to General Assembly resolution 53/35, The Fall of Srebrenica'. See section VIII, especially pages 318–93; *Human Rights Watch*, 1995. 'Bosnia-Herzegovina: The Fall of Srebrenica and the Failure of UN Peacekeeping.' HRW Index No D713.

crimes against their own villages, many of the Muslim defenders of the enclave were themselves victims of earlier Serbian atrocities. Known as *torbari* (the bag people), they had enlisted specifically to exact revenge for past wrongs; an exemplification of the way in which cause and effect become inseparable.[56] In the Yugoslav context, 'barbarization' was compounded and exacerbated by what Alex Danchev calls 'experiential saturation, a compound of conditions and conditioning, nature and nurture'.[57] The Srebrenica massacre was the culmination of a complicated interlinking of cause, effect, and reciprocity, which produced a mix of vengeance and despair.

We can see the conflation of hatred and enmity again during the opening of the Muslim-Croat conflict in 1992. This war within a war, between two ostensibly allied armies was every bit as brutal as the war with the Serbs. It is perhaps symbolized best by the wanton destruction of the medieval Turkish bridge at Mostar.[58] The bridge signified, quite literally, the bridge between cultures – between Islam and Christianity. Although it had survived for centuries (including two world wars), this emotive piece of iconography was destroyed by Croatian artillery specifically because of its iconic status. Like the Srebrenica genocide, the siege of Sarajevo, and the prisoner of war camps, the destruction of Mostar's iconic bridge is one of the most vivid ocular manifestations of hostility and enmity.[59]

Indeed, if we look at the causes of the Muslim-Croat conflict, we can glean insights into the way in which these explosions of hostility pervaded the whole war; constantly changing, evolving, and, of course, igniting as they reacted with the real world. Despite having a common enemy, these two communities fought a vicious and costly war, providing the Serbs with a year of peace with which to muster and prepare their forces. What caused it? In similarity to Srebrenica, several interwoven elements seem to have provided the basis for hostility. Firstly, the war for control of BiH was largely driven from outside of its borders. Bosnia-Herzegovina had been caught up in the process of dissolution not by its own doing, and by the time Yugoslavia began its disintegration Milošević and Tudjman had already met several times to plan the partition and annexation of BiH. From this evidence it is clear that Croatia had a national goal in BiH of grabbing territory

---

56   Kalyvas and Sambanis, 'Bosnia's Civil War Origins', 191–229.

57   Alex Danchev, 'Review Article: The Hospitality of War', *International Affairs*, 83(5) (September 2007), 964.

58   'The Other War', *Time Magazine*, May 24, 1993; 'As Mostar Bridge Crumbles, So does Bosnian Dream of Ethnic Unity', *The Washington Post*, August 30, 1993. See: Charles R. Shrader, *The Muslim-Croat Civil War in Central Bosnia – A Military History, 1992–1994* (Texas: Texas A.&M. University Press, 2003).

59   The concentration camps uncovered by Roy Gutman were especially provocative because images of starved soldiers invoke memories of World War Two. See: Gutman, Roy, *Witness to Genocide: The First Inside Account of the Horrors of 'Ethnic Cleansing' in Bosnia* (Longmead: Element, 1993).

should this become attainable (Silber and Little 1995: chapter 6).[60] Although in the initial stages of the war Muslim and Croat factions co-operated against the Serbs, a combination of outside pressure and the cause-and-effect maelstrom of war seem to have created the perfect environment for conflict between the combatants. The Croat leadership in BiH was politically and materially dependent on Zagreb and was therefore promoting the Croatian national program at the expense of joint Bosniak-Croat cooperation. Further tension was exacerbated and the conflict ignited when large numbers of Muslim refugees fleeing successful Serbian offensives in the early part of the war sought sanctuary in Croatian areas. The influx of refugees caused acute societal tensions between Muslims and Croats, and these fears were easily fostered by Croat politicians with an eye on territorial expansion.[61] The mixture of external policy and aggression, the ethnic cleansing carried out by Serb forces and the tension inflamed by large numbers of refugees relocating to ethnically Croatian areas all converged, igniting and inflaming this conflict. Although Croat-Muslim rapprochement eventually resulted in operational success against the Bosnian and Krajina Serbs, the relationship remains bitter, and is fraught with suspicion and hostility long after the war has ended.[62]

This phase of fighting, and the motivations that led to it, demonstrate the tangled array of antagonisms which can conflate, and which can bestow war with a momentum all of its own. Riven by ethnic suspicion and growing inter-communal hostility, the HVO and ARBiH schism provided the Serbs with the time and space with which to cement their territorial gains without hindrance. Even more worrying for the Bosnian Muslims, they were now confronted with not one, but two enemies.[63]

---

60   During the early weeks of fighting in Croatia, Tudjman had confidentially mooted the idea that the Bosnian TO mobilize in defence of Croatia, opening up a second front against the JNA. Izetbegović would later declare that 'this is not our war' when the JNA began mobilizing Bosnia citizens. However, the Bosnian government did allow the JNA to use the republic as a transit route for JNA operations in Croatia. See: Silber & Little, *The Death of Yugoslavia*, 291–2. See Chapter 6 for an examination of Croatian policy regarding Bosnia.

61   Ibid., 294; Marko Attila Hoare, *How Bosnia Armed* (London: Saqi Books, 2004), 81–7.

62   Steven Erlanger, 'The Dayton Accords: A Status Report', *The New York Times*, June 10, 1996; David Chandler, *Bosnia: Faking Democracy after Dayton* (London: Pluto Press, 1999).

63   The situation worsened in the winter of 1992–93. This was aggravated further by the Vance-Owen Plan, which legitimated Croatian policy. Designed to divide Bosnia into cantons, most of the Croatian majority cantons were positioned next to the Croatian border. Effectively this facilitated Croatian war aims with international legitimisation. Owen later claimed that this argument was 'quite wrong'. Nevertheless, while Owen was right that ethnic tension and hostility was already evident, it remains the case that the Vance-Owen Peace Plan (VOPP) exacerbated an already delicate situation and played into Croat hands.

It is abundantly clear that hostility engulfed the wars in Croatia and Bosnia, as it must to some degree all conflicts. Although one can reject the notion that these wars stemmed from a regional propensity to quarrel – the so-called ancient ethnic-hatreds thesis, existing anxieties were increasingly expressed in ethnic and cultural language prior to the outbreak of war. As the work of scholars such as Susan Woodward has demonstrated, this was the result of political breakdown and transition combining and conflating with the structural faults in the Yugoslav system. Nevertheless, when war occurred these anxieties were intensified by cultural memory and political engineering. Not least, ethnically engineered nationalism was an important engine for recruitment. As the different stages of the war melded into a wider conflagration, pre-existing hatreds became interwoven with the war as it evolved. A mixture of settling scores, opportunism, and revenge, all conspired to underpin and accentuate old tensions. However, it seems that it was the enmity released by the conflict rather than the pre-existing antagonism that produced the most explosive manifestations of violence and hostility.

Passion

As Clausewitz warns, war is fought against an animate object which 'reacts'. A key determinant in winning or losing, passion can be the key to victory, or the portent of defeat. If passions are strong, unleashed as they were in the Napoleonic era by way of the *levée en mass*, this can augur well for victory. It was the national fervour encapsulated in the French military structures which brought Clausewitz to associate the first component of the Trinity with the 'people'. In the new age of warfare which Clausewitz observed, 'The heart and temper of a nation' was the arbiter of whether war would unleash uncontrollable passions. The nation or state with an abundance of passion would have a significant advantage over its challengers. Conversely, of course, the opposite is also true. Clausewitz remarks,

> Loss of moral equilibrium must not be underestimated merely because it has no absolute value and does not always show up in the final balance. It can attain such massive proportions that it overpowers everything by its irresistible force. For this reason, it may in itself become a main objective of the action …[64]

As non-linearity suggests, passion can be somewhat difficult to pin down and can manifest itself in myriad ways. In Clausewitz's own words, 'the strength of the will is much less easy to determine and can only be gauged approximately'.[65] Nevertheless, using the war itself as a benchmark, we are presented with several

---

Owen, *Balkan Odyssey*, 59. On the plan itself, see ibid., 89–126. For further information, see Silber and Little, *The Death of Yugoslavia*, 295, Hoare, *How Bosnia Armed*, 94–7.

64   Clausewitz, *On War*, 275.
65   Clausewitz, *On War*, 77.

clear examples where the evaluation of passion is possible. Starting first with the Croatian example, this section then goes on to explore the extent to which passion influenced the Bosnian Muslim war effort. It then contrasts these examples against the Serbian experience.

Croatia: War and Hostility – The Battle for Vukovar

Until September 1991, the Croats had been disorganized, had limited numbers of trained soldiers, and had been hampered by the political immaturity of their leaders.[66] It was September before the government eventually set up the Croatian Army's general staff, installing General Anton Tus as Chief of Staff. As Tus recalls, 'by that time the JNA had already occupied a quarter of the territory of Croatia'.[67] Prior to this, the Croat forces were left to react to Serb advances and cope as best they could. Tus calls this phase 'disorganized defence'.[68] With the JNA openly assisting the Serb irregulars, the Croats could do little else. From the early stages of the conflict, which included the annexation of territory and the forced expulsion of the Croat population, to the sieges of Vukovar and Dubrovnik, the influence of hostility was widespread and exhibits the nexus between escalation and reciprocity which underpins and drives real war.[69] It is this 'interaction of opposites' which Clausewitz argued so complicates the decision-making process. Despite facing a qualitatively superior opponent, the Croats held firm, eventually obviating the success of Serbian policy in Croatia. The battle for Vukovar provides an important example of how passion influences warfare.

The Serbian assault on Vukovar was ongoing from the very beginning of the conflict. The surrounding towns and villages had already been 'cleansed' of their Croatian populations. Aiding Serbian irregulars, the JNA began operational

---

66   By the end of May 1991, Croatia had several units operating within its borders. This included a reorganization of the Ministry of the Interior (MUP), and the formation of its anti-terrorist units. Together with the remnants of the TO, this formed the basis of the Croatian Army. Not unsurprisingly given the political environment in which it was born, it was beset by disorganization and a lack of supplies. See: Anton Tus, 'The War up To The Sarajevo Ceasefire', in Magaš and Žanić, *The War in Croatia and Bosnia-Herzegovina* (London: Frank Cass, 2001), 48. Especially in the initial stages of the conflict, the Croatian Army had to rely on home-made weapons in order to resist the JNA. Croatia to step up weapons production', *Financial Times*, 10 September 1991; 'Croatia battles to arm its soldiers', *The Independent*, 17 September 1991.

67   Tus, 'The War up To The Sarajevo Ceasefire', 48.

68   Tus, 'The War up To The Sarajevo Ceasefire', 46.

69   Although the battle for Vukovar was bloodier, the international attention focused on Dubrovnik, and drew international condemnation of the Serbian tactics. 'Yugoslav's army cuts off Dubrovnik, *Financial Times*, 3 October 1991; 'Federal War Machine Bears Down on Splendours of Dubrovnik'; and 'Plea for Help as Yugoslav Guns Shell Dubrovnik', *The Times*, 3 October 1991; 'Dubrovnik penned in a ring of fire', *The Independent*, 4 October 1991.

deployments in Baranja and Eastern Srijem. When the build-up of Serbian forces was complete, an offensive against the city began on 24 August 1991. Despite the overwhelming 'conventional' force unleashed against it, the city managed to hold out until mid November. Estimates at this point put the number of casualties at around 2,500–3,000.

The battle for Vukovar was immensely destructive, and was followed by war crimes against those who remained after its capitulation.[70] Yet, it also heralded a turning point in the conflict. After months of bombardment and bitter street-by-street fighting, the JNA and Serbian paramilitary groups eventually took control of the city. It was a bitter blow to the Croats, many of whom had viewed the siege as if it were Croatia's 'Stalingrad'.[71] According to JNA General Panić, the fall of Vukovar had won Serbia the war because the JNA could now march on Zagreb without any interference – the JNA's aim was to take Vukovar and then press on to Zagreb, destroying Croatia's ability to resist in the process. However, although defeat at Vukovar had generated an initial panic in the Croatian press that 'Osijek is finished' (Osijek is the capital of Eastern Slavonia), these feelings began to subside as a knockout blow failed to materialize.[72] In fact, as Norman Cigar has demonstrated, the early Serbian offensives in Croatia reached their culminating point at Vukovar. Serbia's early strategy aimed at seizing territory had proved insufficient for a quick victory.

Although the JNA eventually took the city, the battle had been expensive and had demonstrated that a quick Serb victory was unlikely. In the light of this, General Panić's insistence that Croatia had lost the war looked decidedly less convincing. The initial JNA plan was to separate Croatia into four manageable segments: Gradiška to Virovitica, from Bihać, Karlovac and Zagreb, and from Mostar to Split.[73] This aim, together with the destruction of the Croatian Army, had to be abandoned following the difficulties experienced at Vukovar.

---

70   Mile-Jastreb Deaković, the Commander of Croatian forces in Vukovar estimated that 1,000 people were murdered in one day after the city's capitulation. Transcript of the Interview, Mile-Jastreb Dedaković, 'Croatian commander of Vukovar'. Conducted for the Brook Lapping television series *The Death of Yugoslavia*, 3/17. Transcript dedposited in the LHCMA. See also, 'Hand-to-hand fight for Vukovar', *The Financial Times*, 18 October 1991; 'Vukovar's Wounded Evacuated; Corpses Litter Site of Yugoslav Battle', *The Washington Post*, November 21, 1991.

71   This was the official propaganda line sold by Zagreb. Markus Tanner, *Croatia: A Nation Forged in War*, second edition (New Haven: Yale University Press, 2001), 265. 'Croatians Fight Battle of the Box', *The Telegraph*, 8 September 1991.

72   'Serbian forces pound Osijek', *The Times*, 11 September 1991; *The Independent*, 25 November 1991. Osijek had been on the wish list of the JNA since the first stages of the war; its size and importance for Zagreb assured that it would be a key target.

73   Martin Špegelj, 'The First Phase, 1990–1992: The JNA Prepares for aAgression and Croatia for Defence', in Magaš and Žanić (eds), *The War in Croatia and Bosnia-Herzegovina*, in Branka Magaš and Ivo Žanić (eds), *The War in Croatia and Bosnia-Herzegovina*, 28.

In sharp contradistinction to the JNA, although Croatia's fledgling army was under-resourced, it was welcoming rising numbers of volunteers, many from the Yugoslav army. During autumn 1990, the Croats mobilized 60,000–65,000 soldiers, followed by a second levy of 100,000 personnel in spring 1991. In fact, according to Croat government statements, over 9,000 Serbs actually fought for the Croats in the initial phases of the conflict, and over eighty per cent of Croat officers serving in the JNA deserted as the war began.[74]

Despite defeat at Vukovar, the Croats had become emboldened and had begun to inflict losses on the JNA and Serb irregulars. Although an attempted breakout of Vukovar had been repelled, the Croatian Army had itself repelled major JNA offensives during October. Indeed, although Croatia's early resistance was often ineffectual, it began gaining strength and by October 1991 was in a position to obviate Serbia's strategic plan. The battle for the city is an excellent example of the reciprocity of war at work. Vukovar had been used by the Croatian authorities to inculcate the population with the will to resist and one of the major consequences of the battle was the stiffening of Croatian resolve.[75] In fact, according to Mile Deaković, Commander of Croatian Forces during the battle of Vukovar, the battle for the city served two purposes. Firstly, the destruction of the city identified the aggressor to the outside world, emboldening the Croat population with the grit to repulse the Serbian offensives. Secondly, it was meant to demonstrate Croatian sovereignty. According to Deaković himself, there is some evidence that the Croatian government purposefully restricted the amount of aid reaching the city. According to some sources, munitions, and especially heavy weapons, were not made available despite pleas for them. The argument here is that the Croatian government purposefully used Vukovar as an example of Serbian aggression; it could then be used to draw the international community into the conflict. Although this fits with Zagreb's policy of generating a victim-centred strategy to draw attention to Serbian aggression, the notion that Vukovar was part of that policy has been firmly rejected by Antun Tus.

Whether this is true or not, despite being defeated at Vukovar the Croatian army began to gain the upper hand in the coming months. Dogged resistance had highlighted the weaknesses of the Serbian and JNA forces, both to the Serbs and their adversaries. This directly affected the Serbian decision to open diplomatic channels with the UN regarding the cessation of hostilities in Croatia; although, as will be explained further in the following chapter, the Serbs were quick enough

---

74   Figures cited in: Tus, 'The War up To The Sarajevo Ceasefire', 49.

75   It has been claimed that munitions, especially heavy weapons were not given despite pleas for them. This suggests that it was better for the Croatian government to be viewed as defenders. Transcript of the Interview, Mile-Jastreb Dedaković, 'Croatian commander of Vukovar'. Conducted for the Brook Lapping television series *The Death of Yugoslavia*, 3/17, pp. 2–3 Transcript dedposited in the LHCMA. The initial defence of the town was supplied by an assortment of militias and paramilitary groups as well as the Croatian Army.

to negotiate terms which were favourable to their wider aims. As far as the war in Croatia is concerned, Vukovar's importance as a pivotal event relates to the fact that it proved to be a turning point in the conflict.

For Serbia, the battle of Vukovar was a major strategic setback, precipitating a policy change in Belgrade. The JNA generals may have wished to progress towards Zagreb, but Milošević recognized the futility of further military action. Indeed, as Serbia proceeded towards a peace settlement, it was the Croatian army rather than the JNA or Serb irregulars that were in the military ascendancy. According to Tus, 'they [the Croatian Army] needed only another five to seven days to reach the Sava' (the river bordering Serbia).[76] Despite an apparent qualitative advantage, the Serbian offensives in Slavonia were driven back, and the war took on a new character. The obstinacy, stubbornness, and resolve of the Croatian defenders at Vukovar exemplify the importance of Clausewitz's inclusion of passion as a central element of the nature of war.

Passion, hatred, and enmity are ever present; they maintain and engulf the maelstrom of real war, providing one of the central features of conflict. Although it is relatively straightforward to quantify the numbers of men and guns and come to the conclusion that, on paper at least, the Croats didn't stand a chance, real war is more fickle. War on paper is complicated by the reciprocity of real experience, of fighting against an opponent trying to thwart one's aim. It is the moral forces required to stand firm in the face of adversity and the desire to fight on despite seemingly insurmountable odds, lack of equipment, training or logistics that Clausewitz maintains 'cannot be classified or counted'. The psychological effects of victory, or in the case of Vukovar, stubborn resistance, had cumulative effects which increased morale, and acted as a foundation from which the Croats could take the war to the Serbs. Of course, for the opposing side the opposite is true and the inability to maximize strategic goals has the potential to subvert and diminish morale; as it did in the case of Serbia. It was these elements which Clausewitz believed his contemporaries failed to account for. As Clausewitz observes, 'This type of knowledge cannot be forcibly produced by an apparatus of scientific formulas and mechanics'.

Bosnian Muslim Resistance and the Formation of the ARBiH

The initial stages of the conflict in BiH were similar to those in Croatia. The JNA had begun arming Serbs there as early as 1990, and as the war in Croatia deepened it accelerated the mobilization of Serbs into its ranks.[77] This intensified as the Sarajevo ceasefire brought the fighting in Croatia to an end. The cessation of hostilities also meant that JNA units from Croatia moved into the Bosnian Krajina, Tuzla, Derventa and Brcko. Estimates place the strength of Serbian and

---

76   Tus, 'The War up To The Sarajevo Ceasefire', 64.
77   James Gow, *The Serbian Project and its Adversaries* (London: Hurst & Co., 2003), 172.

JNA forces at approximately 100,000 men and officers, 750 to 800 tanks, 1,000 armoured personnel carriers, mortar and artillery weapons, up to 100 fixed-wing aircraft and as many as 50 helicopters. The position of the Bosnian Muslims was undermined further by the impotence of the Bosnian Parliament and government, which allowed the JNA to maintain the pretence that it was acting to hold the opposing sides apart. The Bosnian President, and leader of the Muslim Party of Democratic Action (SDA), Alija Izetbegović, even asked the army to act as peacekeepers as the tension began to rise in the months prior to Bosnia's declaration of independence. When the JNA ordered the seizure of TO weapons in May 1990, Izetbegović facilitated their removal. In hindsight, the consequences are obvious and when war did break out the Bosnian army was ill prepared and unable to oppose the considerably more powerful JNA. Although the JNA was removed from BiH in May 1992, ostensibly because it was a foreign power, it was actually divided into two, one half – the newly-named VJ – moving into Serbia and the rump Yugoslavia. The other, the VRS, remained in BiH. This newly-named and supposedly independent Bosnian Serb Army was independent only in name and it retained substantial military hardware which provided it with a significant military advantage vis-à-vis Bosniak Muslim and Croat forces.[78]

On 7 April 1992, the same day as Izetbegović declared Bosnian independence, Bosnian Serb leaders issued the proclamation of independence of the 'Serbian Republic of Bosnia-Herzegovina'. Documents seized by the ARBiH demonstrate that the JNA had been actively arming Serb villages and paramilitary organizations, despite the decree that all TOs and paramilitary organizations should return weapons to the JNA in May 1990.[79] Aided by the JNA, the superiority of Serbian firepower meant that the Serbs enjoyed the ascendancy for much of the conflict. The ARBIH and the HVO did attempt offensive actions during 1994, but these were easily repelled by the VRS. It was only in the later stages of the war in 1995, when they were undertaking operations with the Bosnian HVO and the Croatian Army – and backed by NATO airpower – that they were successful against the VRS. By this time the Bosnian Serbs had been isolated by the Milošević regime, were feeling the pinch wrought by international sanctions and were in direct contravention of the international community, suffering from NATO air strikes as a result. However, this did not diminish the tenacity of the Bosnian Muslim defenders.

It is useful to draw a comparison with Croatia's position, especially in the Croatian Krajina during 1991. While the Croats managed to hold on to 15,000 weapons, the Bosnian TO – apart from in Serb dominated areas – lost 300,000 pieces of assorted weaponry. In the initial stages of the war the Bosnian Muslims had no other option but to rely on a disparate range of paramilitary and militia

---

78   The VJ and VRS maintained force strength of approximately 80,000 each. Milan Vego, 'Federal Army Deployments in Bosnia-Herzegovina', *Jane's Intelligence Review* (October 1992), 445.

79   See: Divjak, 'The First Phase, 1992–1993', Gow, *The Serbian Project*, 174.

groups. Although without appropriate equipment, the TO structure, reformed as citizens units, used any means to stem Serbian offensives and a new TO structure, the Territorial Organization of the Republic of Bosnia-Herzegovina (TORBiH), was formed in April 1992. According to figures released by Divjak, around 75,000 volunteers presented themselves to the new TO. It was the TO, and to an even greater extent the quasi-paramilitary force, the Patriotic League, that formed the nucleus of the ARBiH. The Patriotic League had already formed a clandestine military wing prior to the war and this formed the basis for the new army. The Patriotic League joined the TORBiH on April 12.

Although the Bosnian Muslims suffered most from international sanctions prohibiting arms sales to Yugoslav states, they carried on fighting and by the later phases of the war had gained a degree of symmetry with their opponents. By 1994 the ARBiH mobilized an army of 110,000 troops, with 100,000 reserves and a structure of six corps. The VRS, though better off in terms of military hardware, had a force of 80,000, while the HVO was composed of approximately 50,000 troops.[80] Although the ARBiH could not fall back on the resources of a national state in the way that the VRS and HVO could, and despite having a paucity of heavy weapons, they made up for this in the flow of recruits willing to fight – a situation in stark contrast to the Serbs.[81] Despite a clear qualitative disadvantage, the ARBiH held its own, averting the capitulation of Sarajevo and in the process obviating one of the central thrusts of the Serbian project. Figures compiled by the Sarajevo Documentation Centre also illustrate that the ARBiH suffered from higher numbers of casualties than their adversaries. Over the course of the conflict the army lost a total of 30,633, the JNA-VRS lost 20,626, while the HVO had an estimated loss of 5,716.[82] According to Jovan Divjak, seized VRS communiqués suggest that Serb forces intended to take Sarajevo within seven to ten days, and the whole of BiH within three or four months. Their failure to achieve this appears to have had a negative impact on morale, forcing the Serbs to reconsider their operational objectives. Divjak notes,

> The JNA greatly underestimated the human factor. Apart from numerical advantage, the defence enjoyed high morale, had strong patriotic feelings and was able to rely on the population, especially in the larger cities.[83]

---

80   *The Military Balance 1994–1995*, International Institute for Strategic Studies (London: The Institute, 1994), 82–3.

81   The comparative ratio of tanks was ARBiH 30-40 tanks and around 30 APCs. The estimate of VRS hardware was: 330–400 tanks. Ibid., 84–5. This is described by Kalyvas as 'symmetric nonconventional' war. Stathis N. Kalyvas, 'Warfare in Civil Wars', in Isabelle Duyvesteyn and Jan Angstrom (eds), *Rethinking the Nature of War* (Abingdon: Frank Cass, 2005), 88–108.

82   http://www.idc.org.ba/presentation/soldiers.htm (accessed 22 July 2008). Figures compiled by Sarajevo Documentation Centre.

83   Divjak, 'The First Phase', 157.

He continues, 'the (JNA) euphoria began to evaporate, and from then on the enemy attempted to proceed step by step, a little in this direction, a little in that'.[84] Out of a total number of Bosnian Muslim casualties of 45,110 throughout the entire war, 30,442 died during the first year of fighting, an illustration of the severity of the fighting at the beginning of the conflict.[85]

As Clausewitz was well aware, victory or defeat can hang in the balance and the outcome of winning or losing often depends on no more than the vicissitudes of events as they unravel. Unexpected resistance, poor morale, or the loss of an influential backer can complicate the already difficult relationship between purpose and means and these can often add up to reveal unexpected problems. The Muslim government and its forces had to withstand considerable military and political pressure. Not only was it fighting a war on two fronts against the VRS and HVO respectively, it also had to quell internal revolts against renegade factions of the army.[86] More significantly, through the winter of 1993 to the summer months of 1994 it also had to overcome the rebellion of the Muslim population loyal to Fikret Abdić, which declared the area of the Cazinska Krajina the 'Autonomous Province of West Bosnia'.[87] Although the ARBiH did not achieve full operational ability, its very being is testament to the passion of the Muslim defenders and it played a substantial and absolutely critical role in preventing the Serbs from winning the war.

The initial Serb offensive in 1992 had left the core of the Bosnian state surrounded, yet it failed to finish the job. The will to resist was patently an important factor. However, just as we can highlight the role of passion as a component in the Bosniak defence, we must also ask what the impact of Bosniak and Croatian resistance was on Serbian tactics and strategy. It is clear that hostility, passion, a sense of injustice, and not least the visceral and intuitive fight for survival, stirred the ARBIH to continue the fight despite the mounting combat losses and the precarious position of the Izetbegović government. For long periods, the Bosniak forces and government were isolated and as such relied on their own resilience. Despite international recognition, the ARBiH was prohibited from legally rearming because of the international arms embargo imposed on Yugoslavia as a

---

84   Divjak, 'The First Phase', 157.

85   http://www.idc.org.ba/presentation/ethnicity.htm (accessed 22 July 2008). Serb losses for the same period, though much lower, are significant – the Serb casualties during 1992 were 11,157. This can be contrasted with data of Serb casualties for other years: 1993 – 3,731, 1994 – 3,143, and 1995 – 4,970. Bosniak figures are: 1993 – 11,775, 1994 – 5,933, and 1995 – 13,987. Croat figures are: 1992 – 3,243, 1993 – 3,531, 1994 – 488, and 1995 – 357.

86   As mentioned in the introductory chapter, although the new war theorists claim that a central feature of the conflict was organized crime, the ARBiH subsumed these groups as it consolidated its power. See: Hoare, *How Bosnia Armed*, 97.

87   Hoare, *How Bosnia Armed*, 114.

way of preventing the escalation of the war in 1991.[88] The Bosniak forces gained symmetry with their adversaries through a mixture of willpower and numerical advantage.[89] In relation to the will of the ARBiH to resist, Jovan Divjak recalls,

> The aggressor obviously set an unrealistic aim for the occupation of Bosnia-Herzegovina … After initial success in attacks on towns in the Drina valley and the rapid conquest of most of eastern Bosnia, the aggressor was compelled to cut back his unrealistic plans to attainable goals as soon as he came up against the first organized resistance and began to suffer losses.[90]

The JNA and VRS: The Illusion of the Serbian Leviathan?

As October drifted into November (1992), the JNA had attained the majority of its objectives in the war with Croatia. However, by the end of November it was showing signs of weakening at a dramatic rate. Although the JNA's prevalence in conventional weapons, particularly artillery pieces, provided a considerable advantage, it was the paramilitary forces of Arkan's Tigers and Vojislav Šešelj's Chetniks which were consistently used in fire-fights that the regular army was not up to – it was these groups which first entered Vukovar. Why were these forces favoured over the regular army?

By 1988, the JNA had already decided that Serbia offered the best chance of preserving and enhancing Yugoslav centralism. At a conference of the LCY within the JNA, on the 29th and 30th May 1988, it was overwhelmingly decided that the military would opt for ideological centralism. The army's position was at risk if Yugoslavia dissolved; it wanted to maintain its special position in Yugoslav politics. Rooted in the national wars of liberation, the army owed its allegiance to the Federation rather than the republics and it tried to maintain this special position by gravitating towards Serbia's centralist, and later Greater Serbian, vision of Yugoslav politics. That the JNA high command comprised a majority of Serbs who had fought against the Croatian Fascists during World War Two significantly increased the likelihood that it would gravitate towards Serbian policy. In 1990, Serbs made up eighty per cent of the army's officer corps.[91]

---

88   The arms embargo caused fissures to open between European and US policy and the US latterly undertook a policy of 'illegal' re-arming of Bosnian forces, partly as a response to greater military and financial support emanating from Islamic countries. For more information on the Islamic dimension see: Gilles Kepel, *Jihad: The Trail of Political Islam* (London: I.B. Tauris, 2002), 237–53.

89   See: Norman Cigar, *The Right to Self Defence: Thoughts on the Bosnian Arms Embargo*, Institute for European Defence and Strategic Studies, Occasional Paper 63, London, 1995, 8.

90   Divjak, 'The First Phase', 152.

91   *European Weekly*, 30 November–2 December 1990.

As a consequence of the 1974 constitution, the military had been structured with two equal components: the JNA and Territorial Defense (TO) – the TOs were subordinate to the republican leadership rather than the army high command. Concern that the TO could be used as a basis for republican armies was shared throughout the military's leading coterie and the high command had been looking at ways to increase its control since the mid 1980s. Pushing through amendments to the Law on National Defence that would ostensibly 'modernize the army', the JNA reconstituted power, the consequence of which was to further entrench republican suspicion of the army. Effectively this brought the TO structure under the direct command of the JNA, not the republican leadership. As Martin Špegelj (2001: 20) put it: the 'TO forces were now *de facto* subjected to the JNA'. This would prove to be a pyrrhic victory. The JNA initially gained the upper hand, but the army's irascible handling of the whole affair simply widened the growing fissures between it and the republics; pushing it yet further into the hands of Milošević's Serbia. This was all well and fine as long as the state remained extant. However, as the state fragmented, the Serbian and centralist orientation of the army triggered the desertion of large numbers of recruits when war began, the beneficiaries of whom were the nascent republican armies preparing defence against the JNA.

While the majority of JNA forces remained loyal to Serbia and the JNA, substantial numbers had defected or had simply withered away as the war progressed. Conscripts from Kosovo and Bosnia had little reason to fight Croats in a war aimed at securing Serbian hegemony; especially as they might be next on Serbia's hit list. As the war in Slovenia signalled the breakup of the unitary state, the JNA comprised a total force of approximately 169,000 soldiers, with a further 500,000 in reserve, and estimates suggest that it possessed 1,850 main battle tanks, nearly 2,000 pieces of artillery, 100 MiG 21 fast jets and an assemblage of attack and support helicopters. In all, it was a pretty formidable force.[92] However, this quantitative advantage obscures the extent to which its cohesion and utility had eroded as Yugoslavia itself fragmented. Draft dodgers became a real problem for the authorities and only around ten per cent of those called up in Belgrade actually served. As Admiral Stane Brovet put it at the time, 'defeatism is spreading among the people'.[93] In his memoirs, General Kadijević recounts that the failure to mobilize properly cost the Serbs the war'.[94] Even when modified, an up-dated plan could still not fully account for the disastrously low

---

92   *The Military Balance 1991–1992*, International Institute for Strategic Studies (London: The Institute, 1992).

93   S. Ristic, quoting Admiral Stane Brovet, in 'Rani system Hrvatske protiv armije bez podrske' (Croatia's military system against an army lacking support), Narodna armija, Belgrade, 9 November 1991, 10, found in Norman Cigar, 'Serb War Effort and Termination of the War', in B. Magaš and I. Zanić, *The War in Croatia and Bosnia-Herzegovina*, 204.

94   Cited in Špegelj, 'The First Phase', 36.

recruitment rate.[95] In contrast to General Panić's insistence that Croatia had lost the war, Kadijević notes in his memoirs that the JNA was left without appropriate firepower. The JNA's motorized brigades which had been earmarked to advance on Zagreb and Varaždin were unable to fulfil their original task. As Gojko Šušak, the Croatian Defence Minister, remarked at the end of November 1991, 'the army (JNA) was no longer able to control all the territories it had captured … the JNA had reached the end of its tether'.[96]

The result of this for the war aims of Serbia and the Yugoslav army was profound. Not only was the army losing trained men and officers, those who defected were joining 'national' armies defending their homeland. As the prospect of a quick Serbian victory receded, decreasing morale significantly undermined the fighting capabilities of its forces. Whereas Croats and Bosniaks were fighting both to secede and survive, and thus put up stiffer resistance, the Serbian war aims and Serbia's connection with the JNA were vague from the outset. In many ways the marriage of the JNA and Milošević's Serbia should have produced a potent all-Serb military force. In reality, the army's own fragmentation had left huge staffing and skills shortages which would plague its performance throughout the war. More importantly, despite superior capabilities, the JNA did not display the same will to fight. As one study has put it, the JNA 'discovered that it is not possible to make war without the consensus of the entire population'.[97]

From the outbreak of war the JNA suffered from mass desertions which the army failed to stop. National polls carried out in August suggested that 80 per cent of the population wanted to maintain peace at any cost; in further polls, 54 per cent stated that they did not want to fight, and 23.3 per cent claimed that the war in Croatia was not their war. Official statements aimed at inculcating a sense of urgency and national peril fell on deaf ears – the government line that 10,000 Kurds were massing on the Bulgarian border was risible, and met with derision.[98] In the case of the Montenegrin advance on Dubrovnik, the initial hysteria generated by the claim

95   'Conflict puts loyalty of Yugoslav army to test', *The Times*, 3 October 1991; 'Serbs desert army in droves as advance become shambles', *The Sunday Times*, 13 October 1991. Even in the early days of the war, the mobilization rate was only around 25 per cent, reflecting the fact that the JNA was still effectively a Yugoslav, rather than Serbian, army. For the early problems facing the JNA, see: Anton A. Bebler, 'The Yugoslav People's Army and the Fragmentation of a Nation', *Military Review* (August, 1993), 38–51.

96   'The End of the Federal Army? Why Belgrade's Forces are kept in Check in Croatia', *Raids*, January 1992, 28–9. The JNA's mobilization was beset with yet more problems when the President of Bosnia-Herzegovina, Alija Izetbegović, declared that 'this is not our war'; his public denunciation propelling the army into an even deeper staffing crisis.

97   Ofelia Backović, Milos Vasić and Aleksander Vasović, 'Who Wants to be a Soldier? The Call-up Crisis – An Analytical Overview of Media Reports', in Magas and Zanic, *The War in Croatia and Bosnia-Herzegovina*, 329.

98   Backović, Vasić and Vasović, 'Who Wants to be a Soldier? The Call-up Crisis – An Analytical Overview of Media Reports', 333.

that the Croats were gathering on the opposite side of the border dissipated when no opposition was forthcoming. Embarrassed, the Montenegrin Presidency would withdraw troops from the Dubrovnik front when international condemnation attacked the morality of the operation. At Vukovar, soldiers refused to fight until they were told what they were fighting for – no answer was forthcoming. The high rates of desertion and draft dodging found in the Serbian military – a problem that prevented it from fulfilling its war aims in Croatia and BiH – sit uneasily with the thesis that this war was fuelled by an ancient lust for revenge, suggesting that the nature of the conflict is more complicated than the ancient ethnic thesis assumes. Desertion rates were particularly bad in Bosnia; estimates suggest that between 120,000 to 150,000 who were called up dodged the draft, most fleeing abroad. Norman Cigar notes:

> A common assumption in the West was that the Bosnian Serbs were fanatically committed to the nationalist cause and not amenable to rational factors. However, contrary to this 'conventional wisdom', the Bosnian Serbs' willingness to risk death and maiming in combat in pursuit of a Greater Serbia was surprisingly limited.[99]

Both in Croatia under the aegis of the JNA, and then in Bosnia, poor morale and an unwillingness to fight for the ideal of a Greater Serbia resulted in strategic stagnation. As one Serb soldier put it, 'As soon as Serbia itself comes under attack, we will be ready'.[100] That a large proportion of Serbian soldiers could not relate to the war undermined the entire project. This can be juxtaposed against the motivations on the Croat and Bosnian Muslim sides, which were fighting in many respects for their very survival. This strongly suggests that combat motivation is linked to war aims, and this had far reaching consequences for the character of the war.[101] In the case of the Serbs, two things happened. Firstly, in both theatres the Serbs fell back on their heavy weapons. Secondly, with such obvious manpower shortages they had to rely on paramilitary groups. With such a thinly spread force, the long tours of duty perpetuated existing unrest and undermined the fighting capability of the Serb forces still further.[102]

It is quite clear that individual soldiers displayed reservoirs of passion; General Mladić, for example, was known for his resolve and verve and was not easily shaken by the frailties of the Serbian military machine. Yet, when contrasted with

---

99   Cigar, 'Serb War Effort and Termination of the War', 213.

100   Backović, Vasić and Vasović, Who Wants to be a Soldier', 329.

101   During the later stages of operation STORM, Croatian forces stopped short of rolling back the Serb forces deep into Bosnia, despite the fact that the US had suggested that they should. Although there were interlaced reasons for stopping the offensive, one of these was that the Croats had no appetite for fighting for land that was not theirs.

102   It is estimated that by 1994, 46 per cent of the VRS officer corps had fled to Croatia.

the national passions which stiffened the resolve of the Croat and Muslim armies, the Serbs just did not match up. That is not to say that they were not influenced by passion; they were. It may be that this passion was manifested in different ways according to individual context. In some cases of Serbian barbarism, such as Srebrenica, the complex consequences of many interlacing problems resulted in a lack of passion in the wider Serb community which was transferred and seen as desperation by those fighting. All of the forces were influenced in different ways, all were caught up in the maelstrom of war and all committed crimes. In short, hostility in all its forms penetrated the war and generated reciprocal interaction. As the next chapter highlights, all were equally subject to chance.

Chapter 4
'Chance and Uncertainty'

Chance, and the Nature of the Yugoslav Wars – 1991–95

As demonstrated in Chapter 3, we can trace Clausewitz's comprehension of chance to the constantly changing and evolving nature of a particular war as it unfolds. This relates particularly well to the idea that Clausewitzian theory resembles what today is termed non-linear theory. The cause-and-effect nexus which is at the heart of the Trinity sends out 'ripples', which feed into and affect the continuously evolving nature of war. These small 'ripples' have the potential to have big effects. In the previous chapter it was stated that it is impossible to unearth each and every instance of hostility in all its many guises, and then link them so as to identify clear markers. So too is it impossible to chart and trace the pervasive effects of chance on each and every occasion. As Clausewitz concedes, 'issues can be decided by chances and incidents so minute as to figure in histories simply as anecdotes'.[1] This is not to suggest that an evaluation of chance and its impact in particular wars is impossible. It is possible to trace the role and impact of chance, if not in every case then certainly in a sufficient number so as to extrapolate its salient features. Before doing so, however, it is prudent to reiterate once more that in their Clausewitzian form, chance and uncertainty filter into each and every element of war in the same way as they must the entire social world, often with far-reaching effects. Clausewitz writes:

> But in war as in life generally all parts of a whole are interconnected and thus the effects produced, however small their cause, must influence all subsequent military operations and modify their final outcome to some degree, however slight.[2]

As the purpose of the book is to both investigate the validity of the Trinity and assess its place within the rest of Clausewitzian theory, it is helpful to provide a framework from which the chapter can proceed. As explained in Chapter 3, chance and uncertainty are ubiquitous characteristics, but they are also everyday occurrences. By including them in his theory of war, Clausewitz infuses his model with real world characteristics. However, in the Clausewitzian sense, the most pervasive traits of chance manifest themselves in the interaction between

---

1 Carl von Clausewitz (1882), *On War*, transl. by Michael Howard and Peter Paret (New York: Alfred A. Knopf, Everyman's Library edition, 1993), 720.
2 Clausewitz, *On War*, 184.

belligerents. In short, for Clausewitz, chance equals uncertainty. It is in the interaction of opposites where chance rules. This cycle of reciprocity constantly recreates the perfect environment for the invasive aspects of chance to affect war. For Clausewitz, chance is everywhere and impacts on every action and counter-reaction. It may not be the most important determinant in the mind of the strategist, but as noted already, it is a 'player' nonetheless. The subsequent discussion reflects the extent to which chance exists within a reciprocal cause-effect relationship which is non-linear in its effects – what chaos theorists often refer to as the 'butterfly effect'.[3]

Just as the proliferation of hostility produces multiple outcomes, so too does chance pervade everything. This chapter provides a brief exploration of its manifold effects. To achieve this, chance is investigated through the unfolding relationship between the combatants in Yugoslavia and the international community during the period 1991–95. Although the conflict on the ground is equally bound by chance, by providing a frame of reference at the international level it is possible to convey how the pervasive tendencies of uncertainty in war are influenced by complex inter-relationships as it unfolds. The international dimension of the conflict in former Yugoslavia has been chosen because it offers a rich exposition of the reciprocal interdependent features which give life to chance, and which prevented early settlement of the conflict. The current chapter takes up the narrative where the last left off. Although hostility is also non-linear in its effects, one of the major impacts of that element was the way in which it drew in the involvement of the international community.

After providing a framework from which to base the evaluation, the chapter investigates the ways in which the unfolding and non-linear nature of the war provided opportunities and disadvantages to the belligerents. The chapter does not seek to illustrate the final impact of this component on the dynamic of the Trinity, nor on the final outcome of the wars under scrutiny. As Clausewitz reminds us, chance 'weaves its way throughout the length and breadth of the tapestry'. Like the preceding chapter, the purpose of this one is to draw out the salient features of this component in the Trinitarian model, which can then be measured against the final case study – the role of policy.

The Ubiquity of Chance

How does chance penetrate our working environment, how does it shape our lives? In what ways does chance influence our decisions? How does it alter the present? What will this mean for our future? Did chance penetrate the politics of former Yugoslavia and what impact did it have? It is worth ruminating on the effects of chance in a little more detail. For example, in 1966 Alexsander Ranković, considered by many to be heir apparent to Tito, was expelled from the

---

3   See: James Gleick, *Chaos: Making a New Science* (New York: Viking, 1987), 1–8.

socialist party of Yugoslavia for illegally bugging Croatian citizens, including Tito himself. Had he not been caught it is likely that he would have remained a potent force in Yugoslav politics. A hard-line communist and close confidant of Tito, his removal from office also removed a vocal supporter of Serbian interests in the Yugoslav state. If he had remained in office, it is conceivable that the 1974 constitution would have been more favourable to the Serbs, thus preventing the rise of Serbian nationalism in the 1980s. If Ranković had remained in power, could Yugoslavia's dissolution have been prevented? In the same vein, if Mikhail Gorbachev had not come to power in the Soviet Union would the political forces unleashed by his policies of *perestroika* and *glasnost* have been checked? If they had, would the Cold War have continued, as many international relations scholars suggested it must? The end of the East–West confrontation resulted in the loss of Yugoslavia's prestigious status as leader of the non-aligned movement. Even more disastrously for Yugoslavia's long-term future, the end of this conflict brought with it the end of essential foreign financial aid, particularly from the US government. Yugoslavia had been a site of pivotal strategic interest during the Cold War. When that conflict was over, it no longer represented value for money to US taxpayers. The result of this for Yugoslavia was hyper-inflation and economic collapse. Had the Cold War continued, would the loans have continued to flow, propping up the federal Yugoslav government in the process?

Taking this line of reasoning still further, had the Slovene war of succession been obviated successfully, would the rest of Yugoslavia been spared the wars that followed? Had the JNA escalated its efforts, would it have succeeded in holding the unitary state together? At the time, the JNA was caught between the desire to prevent the dissolution of the federal state and thus escalate the conflict, and its preoccupation with the political situation in Croatia and Bosnia. If the Serbs and the JNA had not desisted when they did, Slovene secession may have been prevented. If it had, would the position of the Croatian government have been even more precarious? With 12 per cent of the Croatian population comprising Serbs, the Croats had different problems to Slovenia and they had to some extent been dragged along in Slovenia's slipstream. If the Slovenes were to secede then so would the Croats. Yet, if this avenue had been blocked, would the Croats have thought twice before taking on the might of the JNA? How did chance interplay with the passions and policies of those making and implementing the decision-making process? If not for the whim of history, could Yugoslavia have survived and the war been averted?

Of course, the answers to these questions are specious and the discussion is not intended as an alternative history of what might have been. Nonetheless, it is important to reiterate Clausewitz's argument that events take place as part of the historical process of which we are all part. As Tolstoy said, history is a 'succession of "accidents" whose origins and consequences are, by and large, untraceable and unpredictable'.[4] Uncertainty engulfs the entirety of the human experience. In war,

---

4   Cited from: Isaiah Berlin, *The Hedgehog and the Fox: An Essay on Tolstoy's View of History* (New York: Clarion, 1970), 18–19.

Clausewitz believed that this uncertainty was magnified to such a degree so as to make war 'a game of chance' ... 'wrapped in a fog of greater or lesser uncertainty'.

In the Yugoslav wars of dissolution we can trace this uncertainty throughout the entire conflict. In one such instance, Balkan specialist Lenard Cohen describes how in 1990 Milošević based his early strategies on an incomplete intelligence report prepared by the JNA.[5] The report suggested that although Yugoslavia had lost its strategic importance to the Western powers, the US and its allies would not intercede if the army stepped in to prevent political fragmentation, and ultimately, republican secession. When the then US Secretary of State James Baker visited the region and restated the US policy of maintaining the political status quo, the ruling Serbian coterie accepted this as tacit acceptance that they could halt the independence movements by invoking fear, and ultimately by military aggression.[6] This idea was given further credibility by the early British position, which also favoured the continued unity of the Yugoslav state. As the Under Secretary for Foreign and Commonwealth Affairs told the Commons in June 1991, 'we and our partners have a clear preference for the continuation of a single Yugoslav political entity'.[7] The Serbs and the JNA inferred that they had been given the green light to reconfigure the Yugoslav state by force. They misread and subsequently miscalculated the signals emanating from the international community. Their decision to go to war in Slovenia was based on unreliable intelligence, which was itself the result of changing motivations among the external powers.

We can also see the debilitating effects of friction from the earliest days of the conflict. The previous chapter highlighted the role of Croatian resistance as an exemplar of hostility in action, of how war is fought against an enemy which at every opportunity tries to counteract one's actions. The original Serbian and JNA policy in Croatia was amended because it met with Croatian hostility. Yet 'friction' is clearly visible too. On paper, the JNA was materially and quantitatively superior to the assortment of Croatian police and militias which provided the early defence of the republic. As Clausewitz cautioned it must, real war proved considerably harder than war on paper; 'in war more than anywhere else, things do not turn out

---

5   Lenard J. Cohen, *Serpent in the Bosom: The Rise and Fall of Slobodan Milošević* (Boulder: Westview Press, 2002), 191.

6   Laura Silber and Alan Little, *The Death of Yugoslavia* (London: Penguin, 1995), 150–51. On its own the intelligence report may have been susceptible to critical analysis, yet the signals from the international community embedded a belief in the Serbian leadership that it would have a free hand in determining the future of the state. In one such instance, despite accusations that Milošević had stolen some $1.8 billion from the Yugoslav banking system in order to shore up the decrepit Serbian economy in December 1990, six months later the European Commission President, Jasques Delors agreed to a $4 billion loan. *Tanjug*, 8th January 1991, SWB EE/0967 II; *Financial Times*, 3 June 1991. By that time the entire Yugoslav banking system was in meltdown and the republics were actively disengaging from federal institutions. *Tanjug*, 9th January 1991, SWB EE/0967 B/15.

7   Hansard, 27. 6. 91, cols. 1137-8, cited in Brendan Simms, *Unfinest Hour, Britain and the Destruction of Bosnia* (London: Penguin Books, 2001), 13.

as we expect'.[8] Although Croatian hostility played a central role in the prevention of Serbian war aims, poor morale, desertion, and bad leadership all contributed to reducing the utility of the JNA war machine.

In another example of the detrimental effects of friction and uncertainty, Honig and Both describe the catalogue of errors which contributed to the murder of over 7,000 men and boys at Srebrenica in July 1995. In the days before the massacre it was the opinion of the commanding Dutch officer tasked with protecting the 'safe area' that the Serbs did not mean to capture the enclave.[9] As Honig has since observed, 'within a week, the "safe area" of Srebrenica would no longer exist'.[10] The misinterpretation ultimately resulted in the murder of thousands of Muslim men and boys. Honig recounts the situation shortly before the seizure of the town:

> Lieutenant-Colonel Karremans at this stage still did not believe that the safe area was under serious threat. In his assessment of the situation on Friday evening, he stated that the Serb activities were 'attempts to provoke and intimidate ARBiH (Bosnian Army) and Dutchbat (Dutch battalion)'. He did not expect 'the seizure of Ops and/or parts of the enclave'. The Bosnia Serb Army (BSA) would try 'to neutralise' the Bosnian Army 'in the short term', but, 'due to shortage of infantry', Karremans argued, 'the BSA will not be able to seize the enclave in the short term'. His superiors in Sarajevo and Zagreb accepted this evaluation. The Dutch Ministry of Defence, monitoring the situation anxiously from the Hague, was reassured.[11]

Throughout the night of July 7, Dutchbat was constantly pressurized by Serb forces. Despite signs that the VRS was beginning an offensive to capture the Srebrenica 'safe area', uncertainty provided the fog in which the Serbs were able to carry out their offensive. When the Dutch eventually called for close air support to deter the VRS advance, the request laboured through the complex UN and NATO chain of command. The command structure reflected the inter-agency basis of Operation Deny Flight which had been initiated by UN Security Resolution 816 in 1993. Classified as 'dual-key', the decision to implement air-strikes had to pass through the 'dual' channels of the UN and NATO. The delay in reaching consensus on this type of mission is racked with friction, and the chance to prevent the Serbian annexation of the Srebrenica enclave was lost.[12] The alacrity

---

8   Clausewitz, *On War*, 227.

9   The 'safe area' idea had been hastily constructed to prevent the fall of Srebrenica in 1993, but like other so-called 'safe-areas' there was no real protection and no political will to preserve them.

10   Jan Willem Honig and Norbert Both, *Srebrenica, Record of a War Crime* (London: Penguin Books, 1996), 3.

11   Honig and Both, *Srebrenica, Record of a War Crime*, 9.

12   Alexander Benard, 'Lessons from Iraq and Bosnia on the Theory and Practice of No-fly Zones', *The Journal of Strategic Studies*, 27(3) (September, 2004), 454–78,

required to manage the fluid operational milieu in Bosnia at this time was simply not achievable. As Richard Holbrooke has observed, the dual-key approach was more akin to a 'dual-veto'.[13] Systemic organizational sluggishness prevented the flexibility of response required to react effectively to this type of crisis.

In one example during the debacle, worried that the VRS had acquired a new anti-aircraft capability, NATO warplanes were made to circle the Adriatic rather than Bosnia, in case they should be fired upon by the VRS. When they were called upon, they were too far away to make a difference. Only one sortie managed to get through. It took out one solitary VRS tank; not enough to deter the Serbs into ending their offensive. A second sortie was aborted because the pilots could not identify the 'smoking gun' requirement before having to return to base. In another example, circling warplanes were made to return to base because of bad weather conditions; itself the product of chance. Notwithstanding unfortunate weather, the real failure of the UN and NATO rested in their collective inability to react to the fluidity of a constantly changing situation. Both organizations sought perfect information, which would prove uncontrovertibly that they had taken the correct course of action. Whether the massacre of civilians could have been prevented had the commanders and officials who were entrusted with the UN response acted on the probability that the Serbs were launching a major assault is a moot point. However, it is possible to discern the palpable sense of uncertainty which so paralysed the UN from decisive action. The UN failed to protect Srebrenica not because it did not possess the firepower to deter the VRS, but because friction wore down the utility of the force at its disposal. The Dutch peacekeepers and UN and NATO officials failed, essentially, because they sought concrete evidence of VRS intentions where none was available. As Clausewitz was all too well aware, all information in war is by its nature imperfect.[14]

Although these examples illustrate the ubiquitous characteristics of chance, the following exposition uses the wider participation of the international community as the context for the analysis of chance and uncertainty in the wars in Croatia and Bosnia. This international dimension also provides the strategic frame of reference from which to understand the decision-making processes of the combatants as the war evolved. The involvement of the international community failed to provide an early solution. In fact, in some quarters, its actions, or lack thereof, can be construed as part of the problem. This is not to suggest that the international community did not want the fighting to stop; it did. Rather, because their contribution was an

---

provides a succinct but comprehensive assessment of the potential and limits of air-power as a coercive tool in peace-keeping situations.

13   Richard Holbrooke, *To End a War* (New York: Modern Library, 1999), 63. See also: Jane Boulden, *Peace Enforcement: The United Nations Experience in Congo, Somalia, and Bosnia* (New York: Praeger Publishers, 2001).

14   For a full discussion of the unfolding situation during the days leading to the VRS seizure of Srebrenica, see: Honig, *Srebrenica: Record of a War Crime*, especially chapter 1.

extension of international politics, each state had an eye on its own interests. In terms of achieving a lasting cessation of hostilities, there was consistent failure until the Dayton Agreement brought about an uneasy end to the conflict in 1995. The participation of the international community should not be viewed as sitting outside the nature of the war in a very real way the involvement of external states and international organizations became a tangible part of the fabric of the conflict, and their actions and disagreements fed into the calculus of the combatants. Warren Switzer argues:

> When the combatants perceived a continued reluctance [by the international community] to become involved, they began matching each other, if not in scale, certainly in increasingly barbaric forms and techniques. This was partly because of a perception that the external powers would not forcefully intervene. Or, if intervention did come, the map dividing the territory would probably reflect the battle lines at the conflict's end. Thus the deliberate practice of 'ethnic cleansing' was accelerated to realize revanchist aspirations, to terrorise and burden opponents (real and potential), and to break the will of those who saw their salvation in terms of external intervention.[15]

Chance, Uncertainty, and the International Community

The international response to the Balkan Wars (1991–95) took on different forms and went through several stages throughout the duration of the conflict. Beginning with rhetorical statements about the sanctity of the Yugoslav state, international opinion then became divided as independence for Slovenian and Croatia became a real prospect. In terms of tangible action emanating from the international community, the conflict exhibited not just the brutality of war, but also how problematic it was to attain international consensus. Beginning with the initial moratorium on independence negotiated by the European Community's (EC's) Troika, which brought about a solution to the war in Slovenia in 1991, through to the UN peacekeeping missions in Croatia, and later in Bosnia-Herzegovina, the international community struggled to find a solution to the conflict until firmer US involvement resulted in the Dayton Agreement in 1995. Alongside the various UN missions in Croatia and BiH during the period, several high-level diplomatic missions tried unsuccessfully to negotiate a viable peace settlement. Following on the heels of the success of the EC Troika's conclusion of the short 10-day war in Slovenia, the UN was invited by the combatants to bring about a ceasefire to the war in Croatia, which it successfully achieved when the

---

15   Waren Switzer, 'International Military Responses to the Balkan Wars: Crisis and Analysis', in Magaš and Žanić (eds), *The War in Croatia and Bosnia-Herzegovina*, in Branka Magaš and Ivo Žanić (eds), *The War in Croatia and Bosnia-Herzegovina*, 1991–95 (London: Frank Cass, 2001), 228.

Sarajevo ceasefire was signed in January 1992. However, the Vance Plan, which brought a cessation to the hostilities, negated ongoing diplomatic discussions which aimed to avert a wider conflagration in the region, as well as bringing the war in Croatia to a suitable conclusion. The international community consistently failed to understand that the combatants did not necessarily view international diplomacy in the same way as did the international powers. As Switzer notes, 'To them (the combatants), negotiations were part of the arena of struggle, not a path to peace'.[16] The inability of the international community to form policy consensus actually exacerbated the conflict by multiplying 'uncertainty'.

The international Community met with collective failure in its attempts to bring about peace agreements until 1995. The Vance Plan, which brokered the UN-sponsored ceasefire and the deployment of UN peacekeepers into the Croatian Krajina, only came about after the failure of Lord Carrington's earlier peace initiative in the autumn and winter of 1991. Thus, several high-ranking missions, among them Lord Carrington's Peace Plan, the Vance-Owen Peace Plan (VOPP) in spring 1993, the Owen-Stoltenberg Agreement of September 1993, and the Contact Group plan of July 1994, all met with failure. The reasons for these failures are complex and are interlinked both to the conflict and the wider political environment in the international system in the post-Cold War world. However, in a general way we can directly relate the failure of the international community – broadly defined – to its inability to reach consensus. In this strategic environment the rational course of action for the warring combatants was to protect what you already had; and, if possible, to wring further concessions.

The lack of international success in bringing about a negotiated settlement was a pattern formed at the very beginning of the conflict. We can take Lord Carrington's Peace Initiative as a case in point. Carrington was tasked with bringing the war in Croatia to a successful conclusion as well as preventing the outbreak of war in Bosnia-Herzegovina. To achieve this, the 'Carrington Plan' conceived a reworking of the Yugoslav-wide constitutional arrangements which he deemed could be re-drafted to allow the Federal institutions to remain intact. At the same time it would offer the republics as much or as little independence as they wished. In an attempt to address the concerns of all of the republics, the Carrington Plan even envisaged armed national defence forces in regions where large numbers of minorities existed – this was aimed at the Serbs, who argued that Serb minorities in Croatia and BiH were under threat of the supposedly revanchist *Ustashe* and Muslim fundamentalists.[17] Carrington later explained: 'It seemed to me that the right way to do this was to allow those who wanted to be independent

---

16   Waren Switzer, 'International Military Responses to the Balkan Wars: Crisis and Analysis', 228.

17   The plan was similar to the Izetbegović-Gligorov proposal of 1991 – an 'asymmetric federation' that envisaged Serbia and Montenegro at the centre of the state, and BiH and Macedonia as semi-detached but constituent republics. Croatia and Slovenia would be nominally linked to the new state but would exercise as much sovereignty as they wished.

to be independent and to associate themselves with a central organization as far as they wanted to'.

Of course, the main reason for failure was because it supposed that there was a peace to keep. A key failure of the international community was simply to understand that at this juncture, for the Serbs at least, war offered more fruitful rewards. The international response was based on the notion that a little pressure would bring about an end to the conflict, and that all of the combatants wanted the war to be brought to a conclusion. As highlighted in the previous chapter, one of the factors which contributed to the engagement of the international community was the level of violence beamed into Western homes by the international media. In fact, the escalation of hostility played a large part in engaging the external powers; to such an extent that both Croatia and the Bosnian Muslims used violence perpetrated against them as a strategic means – the visualization of hostility was used as a vehicle with which to involve the international community. This heaped pressure on the external powers and eventually resulted in the condemnation of Serbian war aims and Serbia's isolation from the international community.

Nevertheless, bringing the full weight of the international community to bear on the Serbs and engaging the external powers actively and constructively in the conflict was not always easy. In large part this was because the international community was made up of the sum of its parts – states. Their involvement in Croatia and Bosnia was conditioned by their own domestic and international interests rather than overt humanitarianism. For example, the US response was formulated on the need to protect the status quo, while Germany was motivated by economic interests and may have been exercising its muscles as a demonstration of its new-found unity following the end of the Cold War. In contrast, the British response seems to have been premised on several interlaced issues. On one level, diplomatic involvement provided the opportunity to maintain international standing and credibility in the new post-Cold War security environment where its own previous special position may not in fact have been so special after all. The crisis offered the perfect platform from which to act as a Great Power in European affairs. Notions of great-power status were somewhat tinged by a widely sceptical establishment, which constantly delayed meaningful intervention.

The British government, though reflecting an international opinion that the war should be brought to a conclusion, consistently rebutted calls for tougher action on the grounds that it did not fit with the British national interest. This reflected the British government's mood throughout the conflict. As Douglas Hurd explained during 1993, 'it is in our interests to do our bit, but we should not over pretend, or let rhetoric get in the way of reality'.[18] Douglas Hogg underlined the government's sentiments on the matter when he explained that any concerted effort to stop the fighting by force would require being 'underpinned by national will. I do not believe that will exists in this case'.[19] The French were interested in maintaining

---

18   Cited in Simms, *Unfinest Hour*, 7.
19   Cited in Simms, *Unfinest Hour*, 43.

the status quo, although they too had reason to protect their status as a leading power. With regard to Russia, its own involvement was burdened by uncertainty. It initially advocated the status quo, a reflection of its own fragmentation in the aftermath of the Cold War – the war in Chechnya was an especially sensitive issue. In addition, involvement offered the chance to maintain international standing in spite of deleterious internal problems. China, on the other hand, used its power in the Security Council to assuage international disquiet about its own human rights record. Involvement in the crisis was thus a favourable way of gaining bargaining chips with which to negotiate other regional or international issues.[20] In this way the complexity of the war in Croatia, and especially in Bosnia, was complicated further by the changing relationships and priorities of the external powers, as much as their reaction to the combatants, who could never have factored these complex processes into a rational calculus model. It was such complexity that brought Clausewitz to opine that the direction, and even success, of a war may be down to 'chances and incidents' – the result of intersecting lines – that are out of the control of the combatants themselves, 'so minute as to figure in histories simply as anecdotes'.[21]

It is also important that the international response be measured against the wider changes to the international community brought about by the end of the Cold War. Although it proved premature, Jacques Poos' vatic proclamation that 'This is the hour of Europe' had signalled the EC's willingness to tackle serious security issues in the post-Cold War security environment.[22] The European powers, under the auspices of the EC, should thus take responsibility for the unfolding crisis. Preoccupied with the fall-out from the fragmentation of the Soviet Union, the re-alignment of NATO, and domestic calls to relinquish some control so Europe could take its fair share of the burden of international security management, this prospect mirrored the United States' own interests regarding Yugoslavia. This set the international scene for much of the conflict. US policy vis-à-vis former Yugoslavia was to allow the European powers and the UN to bring about an end to the conflict. As the then Secretary of State, James Baker, frequently remarked during the early months of the conflict, 'the US doesn't have a dog in that fight'.[23] This typified the US position. Although the Europeans wanted to exert greater influence in their own 'backyard', the Yugoslav crisis was too big a test too soon and the absence of US influence undermined European attempts to end the hostilities.

The European Community and the states which comprised it were unable and unwilling to force a solution. Worried about the prospect of the crisis triggering a

---

20   Paul Williams, 'The International Community's Response to the Crisis in Former Yugoslavia', in *The War in Croatia and Bosnia-Herzegovina*, 274.

21   Clausewitz, *On War*, 720.

22   Cited in John Zametica, 'The Yugoslav Conflict', *Adelphi Paper*, 270 (Summer, 1992), 59.

23   Silber and Little, T*he Death of Yugoslavia*, 201.

wider European conflagration, the emphasis of the negotiations was on stopping the crisis, lest it escalate into a general European war. It was about containing the problem rather than resolving it. One flaw evident in each of the seemingly perennial rounds of negotiations was the focus on bringing about an end to the conflict because international prestige was at stake. At the height of the siege of Vukovar, for instance, Douglas Hogg exclaimed to the UK parliament that there had been 'repeated ceasefire violations' – this demonstrated the hopelessness of the situation. In 1994, the then Foreign Secretary, Douglas Hurd argued that 'the only people who can stop the fighting are the people doing the fighting. You have at the moment, alas, three parties in Bosnia, who each of them believe that some military success awaits them'.[24] Hurd's observation was undoubtedly correct, yet he failed to realize that the inaction, or at least semi-involvement, of the external powers, was fuelling the most brutal aspects of the conflict.

The international community was wracked by domestic and international issues which impinged on their collective engagement in the Balkans. In this way, 'uncertainty', 'chance' and 'friction', which resulted from the changes in the international balance of power at the end of the Cold War – the non-linear and fluid nature of the international system, filtered into and conflated the nature of the wars in Croatia, and particularly in Bosnia-Herzegovina. To reiterate the point once more, the strategic frame of reference which complicated the international response to the wars in former Yugoslavia fed into the conflict. Of course, this is exactly what Colin Gray highlights when he talks about the 'sovereignty of context'.[25] Just as understanding that the political, cultural and social contexts of war are pre-requisites to understanding its nature; to these Balkan wars we can add the international context. The combatants were constantly reacting to and shaping the international response. Although this produced opportunities for each of the groups to win important concessions from the external powers, it also produced an acute and unremitting sense of uncertainty. It is axiomatic that the complexity at the heart of this interactive cause-effect nexus had a direct bearing on the war.

As the war in Croatia escalated and looked likely to engulf Bosnia-Herzegovina, we see a direct example of the paralysing lack of international unity over the unfolding crisis – the Carrington Plan was usurped by two alternative international initiatives. The first of these, the Vance Plan, fell under the auspices of the UN – it was the Vance Plan which brought a cessation of hostilities in Croatia in January 1992. The other was a German initiative, and exemplified a renewed German foreign policy contribution to international diplomacy. Following fast on the heels of the stuttering Carrington Plan, Cyrus Vance's role as the UN's chief negotiator

---

24   'Eye of the storm' (interview), *Crossbow* (magazine of the Conservative Bow Group), February 1994, 4. Cited in Simms, *Unfinest Hour*, 26.

25   Colin S. Gray, *Another Bloody Century* (London: Weidenfeld & Nicolson, 2005), 55–97.

was anticipated to complement Lord Carrington's ongoing negotiations.[26] In fact, the two plans differed considerably. The most important distinction was that the Vance plan had the option of contributing UN peacekeepers as a way of bringing a cessation to the conflict in Croatia. By offering guarantees for minority rights in the republics and thus maintaining a loose federal Yugoslav state, Carrington's proposal intended to bring an end to hostilities and prevent the conflict from spilling over into BiH. The German proposal, which campaigned for the early recognition of Slovenia and Croatia, would effectively make Carrington's initiative obsolete. For the German government, which had taken a pro-Croat line well before the beginning of the conflict, the Carrington proposals would legally permit a continued JNA presence in Croatia. If Croatia was recognized by the international community, in the eyes of international law the JNA would be an occupying army. Recognition would thus presage the removal of JNA and Serb irregulars, in turn bringing about an end to the conflict. The problem with having such an array of initiatives was that it offered the combatants different options to choose from, consequently making the chances of reaching consensus even smaller – each side simply opted for their favourite choice. In fact, as Laura Silber and Allan Little put it, 'this was international mediation *à la carte*'.[27]

Determined that the German option would be accepted, German Foreign Minister Hans Dietrich Genscher made it a priority of the new German foreign policy.[28] If Germany's European partners failed to recognize Slovenia and Croatia, Germany threatened to do so alone; thus putting the prospect of further European integration at risk.[29] Preoccupied with developing a coherent European defence and security policy, the other EC states accepted Germany's ultimatum. This consequently also resulted in a negation of the Carrington Plan. The problem with early recognition was that it made the prospect of war in BiH more immediate than ever.[30] This had been the view of the then UN Secretary General, who in May 1991 had underlined his personal reservations vis-à-vis recognition to the Dutch Foreign Minister noting that: 'I am deeply worried that any early selective recognition could widen the present conflict and fuel an

---

26   Silber and Little, *The Death of Yugoslavia*, 196.

27   Silber and Little, *The Death of Yugoslavia*, 197.

28   Tanner, *Croatia, A Nation Forged in War.*

29   German involvement in Balkan affairs revivified a deep-seated animus, which was rooted in the Yugoslav experience of World War Two. Serb propaganda consistently propounded the idea that Germany was again flexing its muscles. As Yugoslavia teetered on the edge of war, Croatia's renewed ties with Germany were used to rekindle a fear of a resurgent Ustashe – the war-time Nazi puppet government in Croatia. Claims of German revanchism were increasingly met with international ridicule, isolating the Serbs as a result.

30   Although the EC agreed to consider all applications submitted by 24 December 1991, to then be considered by the Badinter Commission, Germany issued a statement that it would recognize Croatia regardless of the Commission's findings. Although the US withheld its own recognition, it did not intercede to derail German or European recognition.

explosive situation, especially in Bosnia …' (cited in Trifunovska 1994: 428). Lord Carrington's exasperation was palpable. As he later explained:

> It would make no sense at all … If they recognised Croatia and Slovenia then they would have to ask all the other parties whether they wanted their independence. And if they asked the Bosnians whether they wanted independence, they inevitably would have to say yes, and this would mean a civil war (in Bosnia).

Of course, recognition made it more likely that the Vance Plan would be agreed, which it was through the Sarajevo Agreement signed in January 1992. However, rather than bringing finality to the war, the combination of the Vance Plan and recognition triggered a wider conflagration which may have been preventable. The Serbs did concur with the Vance Plan. Retrospectively, however, it is evident that Milošević had already determined that the war in Croatia should be brought to a conclusion. By the winter of 1992, it was clear that Croatian resistance was stiffening and that prolonging the conflict indefinitely risked a reversal of fortunes; the JNA and Serb forces had gained as much territory as was possible. Not wanting to get bogged down in an unprofitable war in Croatia, Milošević looked for ways to extract his forces. The Vance Plan offered the perfect solution. Serbia could redeploy its forces for the coming war in Bosnia without worrying about the security of Serb-occupied Croatia. Integral to the agreement was the inclusion of UN troops in Croatia; these would police the occupied areas of the republic, and by extension maintain Milošević's gains without the aid of the JNA.

Although the Vance Plan effectively ended the war in Croatia, war was quick to move into Bosnia. Reactive rather than pro-active, the international response was plagued by a lack of consistency throughout the war, meaning that the combatants were able to gauge the weaknesses of international resolve and exploit this for their own strategic interests. The same problems were clearly evident throughout the Contact Group negotiations. Like previous rounds of negotiations, the Contact Group plan envisaged the return of territory to correspond to a pre-war demographic position – it proposed the separation of Bosnia into two mini-states, the Muslim-Croat Federation, which would receive 51 per cent of the territory, and Republika Srpska, which would receive the remainder. The architects of the plan again failed to accurately gauge the resolve of the combatants to get what they wanted. Like other rounds of negotiations, the involvement of the external powers reflected a desire for international prestige – to be seen to be doing something was a mark of standing. Finding a meaningful solution came second to international positioning in the new security environment following the end of the Cold War.

More importantly, the failure to find a consensus created a political vacuum which was exploited by the combatants – particularly the Bosnian Serbs and their patrons in Belgrade. Although the international community was united in its condemnation of the apparent intransigence of the combatants, it failed to agree on a coherent policy which could bring about an end to the conflict. Debates about the use of force were particularly contentious and ranged from disagreements on the use of air power to the lifting of the arms embargo, which had effectively paralysed the Bosnian Muslim forces from repelling the Serbs.[31]

This lack of cohesion is again evident when assessing the response to the Markale market bombing in Sarajevo, in which the market was shelled by the VRS with the loss of 66 people on 5 February 1994.[32] The initial international outcry was met with tacit approval by NATO that air strikes would follow against Serb positions if they failed to comply with a NATO ultimatum. The Serbs were ordered to withdraw heavy weapons to a distance of 20 kilometres from the city centre, or place weapons at specified sites under the jurisdiction of the UN. They had 10 days to comply with the ultimatum. If they failed to do so, NATO would use air strikes against them for the first time in the history of the alliance. This time the opposition to NATO airstrikes came from Lord Owen, in his capacity as chief UN negotiator. Owen was concerned that Russia would react badly to the unilateral use of NATO air power. If NATO acted outside the UN mandate, it had the potential to risk East-West relations and, so Owen believed, even trigger a new Cold War. Agreement on the use of NATO met a further blow when the UK questioned the efficacy of air strikes. UK objections were overruled at a meeting of the North Atlantic Council on February 7 and NATO proceeded to make its ultimatum to the Bosnian Serbs.

Concurrently, however, British Lt. General, Michael Rose, working as commander of UN forces in BiH, was attempting equally hard to prevent NATO airstrikes, should they derail the UN sponsored peace process being negotiated by Owen and Stoltenberg.[33] Rose's alternative to NATO was the UN's 'Four-Point Plan', which suggested: (i) an immediate ceasefire, (ii) the withdrawal of heavy weapons to at least 20 kilometres from Sarajevo, or their surrender to UN control, (iii) the interposing of UN troops between the two front lines, and (iv) the establishment of a joint committee to agree the details of the Plan's implementation. Meeting considerable resistance from the Bosnian Serb and Muslim governments, the Serbs only agreed to the proposal when the international fallout over the NATO ultimatum resulted in renewed Russian participation in the

---

31 Although the embargo divided international opinion, it did eventually help to produce a military stalemate. The Serbs had a preponderance of heavy weapons which helped them to carve off large areas of territory during the early stages of the conflict. The Bosnian Muslims had few weapons but a ready supply of recruits. Although the ARBiH was initially out-gunned, as the paralyzing effects of poor Serb morale were conflated with an equally paralyzing paucity of military supplies, the Muslim forces increasingly held parity with the VRS. Although the embargo also covered Croatia, it was able to get access to military hardware, and was thus able to bolster the firepower of the regular army.

32 The Serbs accused the Bosnian Muslim government of shelling its own people in a crass attempt to censure the Serbs and trigger international condemnation of the VRS. The subsequent UN investigations pointed the finger at the Serbs and the loss of life was attributed to the VRS.

33 Rose was commander of UN forces in Bosnia-Herzegovina between January 1994 and January 1995.

mediation process. The result of this was the dispatch of four hundred Russian peacekeepers to uphold Rose's 'Four-Point Plan'. The NATO ultimatum was meant as a show of international commitment and strength, the intention of which was to force the Serbs to the negotiating table. In effect, the agreement reached by Rose not only complied with NATO's ultimatum, it officially partitioned Sarajevo into Serb and Muslim areas, which would be policed by the UN.

In other debates regarding the use of force, national disagreements limited the effectiveness of that involvement. Division was particularly evident regarding the use of air-power. Although Bosnian Serb obstinacy steadily situated them in the role of aggressor, the UN mandate restricted the organization to a purely peacekeeping role. It did not have political authority, nor was it equipped to act as peace-maker. It also worried that changing the name of the game in mid-operation would risk the moral authority of the UN and unduly risk the safety of its personnel. If the advocates of air-power had their way, UNPROFOR would be forced into losing its neutrality. Under-equipped for this change in role, it was feared by many that crossing the so-called Mogadishu line would leave NATO troops susceptible to reprisals should the alliance bomb Serb targets.[34] In fact, when NATO did use air power in a half-hearted attempt to stave off the Serbian assault on Goražde in April 1994, General Mladić took 150 UN personnel hostage as human shields. In a public announcement regarding the Bosnian Serb offensive against the UN 'safe' area of Goražde, US Chief-of-Staff General John Shalikashvili pronounced that NATO would not issue a 'Sarajevo' style ultimatum to the Serbs. Shalikashvili's declaration resulted in an intensification of the Serb offensive. Yet again, the lack of a clear policy from the international community allowed the Serbs to take the initiative.[35]

As the attack on Goražde intensified, on 10 April Rose eventually sanctioned the use of air strikes. NATO's first aggressive operation since the inception of the organization in 1949 saw US Air Force F-16s drop three bombs on Serb artillery positions; a further strike targeted tanks and armoured personnel carriers the following day. A subsequent wave of strikes was then hampered by bad weather; the next suffered the loss of a British Sea Harrier – struck by a VRS surface-to-air

---

34   The Mogadishu line refers to the ill-fated US-led intervention in Somalia in 1993, in which US policy-makers crossed from peace-keeping to peace-making and enforcement, resulting in highly publicized casualties. Although the prospect of western soldiers being used as human shields appears to have dominated discussions, the primary job of UNPROFOR, which was to provide humanitarian relief, would also have been put in jeopardy. Like other humanitarian missions past and since, the UN has to rely on the goodwill of the belligerents if it is to get aid to the people who need it most. If the UN took sides, then the Serbs had the potential to cut off humanitarian relief, which it could argue was sustaining its enemies. During the Srebrenica crisis, Dutchbat feared losing Serbian trust if the UN used military force, and consequently humanitarian aid being cut off.

35   One of the most divisive moments for the external involvement came following the allegation that the French military had directly negotiated with the Bosnian Serbs. It was alleged that the French would veto all requests to use air power if the Serbs vowed not to use French service personnel as human shields.

missile. This incident concluded NATO airstrikes; just as in Srebrenica and Žepa, NATO's commitment to deter the Serbs was dealt a blow by complicated rules of engagement and internal fissures within the alliance. The Bosnian Serbs took full advantage. With NATO deliberating on how best to deal with the hostage situation, Mladić used the opportunity to widen his offensive, attacking the Bosnian-government-held city of Tuzla in the North of the country. Yet again, the lack of a coordinated international response had provided the Serbs with an opportunity to seize the initiative.

Chance, Non-linearity, and the 'Ripple' Effect

The above discussion is not intended as a way of apportioning blame as to the length or intensity of the wars in Croatia and Bosnia. Its purpose is to illustrate how widespread uncertainty results in war being pervaded by chance. As we know, Clausewitz's comprehension of chance and its place within the Trinity is intended to exemplify the unfathomable – war 'wrapped in a fog of greater or lesser uncertainty'.[36] We can sense that uncertainty by assessing the participation and interaction of the international community and its own involvement with the combatants. The continual interaction of all of the actors provided opportunities, if only the combatants could make the right decisions. However, 'with chance working everywhere, the commander continually finds that things are not as he expected'.[37] For Clausewitz, only a 'genius' had the ability to comprehend the array of advantages and disadvantages, and make the correct decision. This is clearly not an easy thing to do; yet, whether one describes this as genius, or simply as yet another example of chance, it is useful to explore some of the key decisions made by the combatants as the war unfolded. It should be remembered that, in the Clausewitzian sense, war takes place in the context of constant interaction between hostility, chance and reason; and 'chance' itself very much reflects the notion of non-linearity.

Throughout much of the conflict in Croatia and Bosnia, it was the Serbs rather than their adversaries who were best able to take advantage of the unfolding drama. They quite correctly inferred that the international community was unwilling to commit itself in a manner which may escalate the conflict or risk national contingents of troops. They enjoyed rich rewards as a result. However, the long-term victim-centred strategy employed by the Croats and Bosniaks in order to provoke international action against the Serbs, and the Serbs' obduracy in the face of international criticisms, gradually isolated the Serbs.[38]

---

36   Clausewitz, *On War*, 117.

37   Clausewitz, *On War*, 117.

38   For long periods of the conflict, UK opinion was sympathetic to the Serbs and attempts to apportion blame were regularly rejected. The comments of Sir Nicholas Bonsor, the then Under-Secretary of State at the Foreign and Commonwealth office, was indicative of UK Government opinion even after the Srebrenica massacre. As he put it, 'this is not

For long periods of the conflict, the Serbs' obstreperous approach to the diplomatic endeavours of the external powers made sense. As some of the examples highlighted above testify, they were quick to seize on international disunity and work it to their own advantage. For example, reflecting on the Vance Plan which brought about the end of the war in Croatia, Borislav Jović later revealed that 'at that point the war in Croatia was under control … Slobodan and I after many conversations decided now was the time to get the UN troops into Croatia to protect the Serbs there'. He continued, 'when Croatia would be recognized, which we realized would happen, the JNA would be regarded as a foreign army invading another country. So we had better get the UN troops in early to protect the Serbs'.[39] The Serbs simply used the UN to police the 'Serbian Krajina', leaving themselves free to concentrate on the unfolding crisis in Bosnia.

Similarly, the failure of the UN and NATO to have a common policy regarding airstrikes left weaknesses to be exploited, which the Serbs did. The NATO ultimatum and General Rose's 'Four-Point Plan' were intended as a resolute stand which would subdue the Serb threat, thus diminishing the preponderance of their artillery power. Yet the failure to reach a consensus provided an opportunity which Radovan Karadžić gladly accepted. The Serbs agreed to the provisions of Rose's plan, prevented a NATO attack, and brought the Russians back into the fray – on their side. More importantly still, they acquired the de facto partition of Sarajevo, a long-term war aim. Silber and Little sum up the situation perfectly:

> Thus did Radovan Karadžić play a bad hand very well. He had been seen, by the international community, to compromise on weapons withdraw. His guns had fallen silent; the killing in Sarajevo stopped. The partition he so badly wanted was beginning to take real shape, and he did not even have to supply troops to defend the urban frontier of his state – the UN was doing it for him. The Bosnian government felt out-manoeuvred and humiliated. Karadžić incredibly though it seemed, had emerged as the principle beneficiary of the NATO ultimatum to use force against him.[40]

This is a wonderful example of the unexpected consequences of actions having the wrong effect, and reflects Clausewitz's identification of the opportunities of reciprocity. Using another example, as highlighted above, when NATO did use force to stave off the Serb attack on Goradže, US General Shalikashvili made a public declaration that NATO would not issue a 'Sarajevo' ultimatum. With immediate freedom from air strikes, Mladić simply escalated his offensive against

---

a one-sided conflict in which there are white hats and black hats at war. It is a conflict in which the depth of bestiality is incomprehensible in a civilized world and it is not confined wholly to the Serbs' (cited in Simms 2001: 26).

39   Transcript of Interview, Borislav Jović, President of the Yugoslav Collective Presidency 1990–91. 3/35 (July 1994–July 1995).

40   Silber and Little, *The Death of Yugoslavia*, 318.

Tuzla. These examples are by no means exhaustive and they are indicative of the way in which at different times the JNA and then VRS were able to out-manoeuvre the entire international community.

Nevertheless, perhaps owing to their preponderance of heavy weaponry, and certainly buoyed by their early successes, the Serbs failed to discern the subtle signs that the tide was turning against them. Chance is apt to rebound; the run of events produced by chance may bring great rewards, but these will even out over time.[41] This is not always easy to identify when war is ongoing and, as Clausewitz tells us, the 'commander' may often believe that 'he will be able to reverse his fortunes just once more'.[42] The problem with such a view is of course axiomatic, being seduced by the prospect of overturning lost gains, the commander or leadership will often try to ride out the storm in a futile attempt to wait for some mercurial change in fortune. Of course, the difficulties confronting political and military leadership are capacious. This is precisely why Clausewitz believed intuition and genius so important. While Clausewitz's contemporary and rival, Jomini, believed war to be like a game of chess, or Sun Tzu like a game of *go*, Clausewitz dismisses anything which attaches immutable rules; it is the unfathomable and chaotic exemplars of chance and complexity that make war more like a 'gamble'; like a game of cards.[43] Although Clausewitz expects the commander or leadership to act on probabilities, thus demonstrating individual skill and dexterity, by their very nature probabilities are far from certain. It is therefore extremely difficult to gauge when the correct decision is being taken. This is why Clausewitz incorporates this everyday element into his theory of war – it may be normal, but is inescapable nonetheless.

Although international inaction emboldened the Serbs, the levels of violence perpetrated by them gradually fostered consensus between the external powers, and by the end of 1994 it was widely accepted that they were the principal aggressors. The Serbs had used brutal tactics from the outset, in the Krajina, Slavonia, and in the Bosnian towns of Bilejina and Zvornik. After the seizure of the Srebrenica enclave in July 1995, world opinion eventually hardened. Although the ubiquity of uncertainty which engulfed the conflict had provided the Serbs with a strategic context in which to maximize their qualitative military advantage, ultimately they failed to recognize that 'chance' has a tendency to work in the opposite direction. Although the Croats and Bosnian Muslims tried to draw the international powers into the conflict so as to ameliorate an often shaky strategic position, there appears to have been an acceptance that the participation of the

---

41   See: Katherine L. Herbig, 'Chance and Uncertainty in *On War*', in Michael Handel (ed.), *Clausewitz and Modern Strategy*, 95–116.

42   Clausewitz, *On War*, 250.

43   Handel, *Masters of War*, 240. For more information see also: Michael I. Handel (ed.), *Intelligence and Military Operations* (London: Cass, 1990), 13–21.

external powers was essentially incongruous – favouring containment rather than overt military intervention against any one side.[44]

Despite this, the non-compliant and refractory behaviour exhibited by the Serbs in much of their dealings with diplomatic and military missions was suggestive of a more sinister anomie, perhaps inspired by cultural and ethnic hatred. This intransigence consistently pitted them against international opinion as well as their immediate opponents, and even caused a split within the wider Serbian project. Anxious that sanctions against Serbia were beginning to bite and therefore risking not only the dream of a Greater Serbia, but also of his own position as Serbian President, Milošević put enormous pressure on the Bosnian Serb leadership to negotiate the termination of the war. Buoyed by a series of easy operational victories in the early years of the conflict, the Bosnian Serbs were in no mood to capitulate their prize. In fact, if one was to place oneself in the shoes of the Serb leadership in BiH, there would be little incentive to comply with international efforts to end the conflict on a model which would see the return of territory already safely in the hands of the VRS. The Contact Group Plan in 1994 is a case in point. The Serbs were expected to surrender thirteen towns, and considerable chunks of territory won in 1992. Yet at this juncture the Serbs held seventy per cent of the territory in BiH. Furthermore, the eastern Muslim enclaves of Srebrencia, Goradže and Žepa would remain in Muslim control – they would therefore also remain a strategic threat to the Serbs. The Republika Srpska Assembly at Pale rejected the plan, precipitating the split with Belgrade. The schism revolving around the Contact Group Plan mirrored earlier disagreements regarding the negotiations over the Vance-Owen Peace Plan and then the Owen-Stoltenberg plan. Following the Serb offensive at Srebrenica in 1993, the UN Security Council passed a resolution which tightened the sanctions on Serbia. This ordered Serb assets to be frozen abroad and transhipments through Yugoslavia were banned.[45] According to Lord Owen, this brought Milošević to the decision that the war should be brought to a conclusion.[46] Owen argues that this was an economic decision on Milošević's part and that international pressure was making an impact. Despite this, Milošević found it impossible to rein in his formal proxies until combined NATO, Croatian, and Bosnian Muslim operations began to force the Bosnian Serbs back into Belgrade's political ambit. As Clausewitz would put it, 'in war more than anywhere else, things do not turn out as we expect'.[47]

Confronted with an unfolding crisis in Yugoslavia, a combination of factors resulted in a hesitant international response. Retrospectively, there was little to prepare the international community for such a crisis, and the difficult balancing act between domestic and international politics impinged on the utility of collective international resolve. The early anxiety about the war escalating should

44   Switzer, 'International Military Responses to the Balkan Wars', 289.
45   Silber and Little, *The Death of Yugoslavia*, 276.
46   David Owen, *Balkan Odyssey* (New York: Harcourt Brace & Company, 1995), 151.
47   Clausewitz, *On War*, 227.

the international community lean too firmly on the belligerents was palpable, and the US in particular had reservations post-Vietnam about being dragged into an unwinnable war. The Serbs did their best to foster this concern. In fact, the principal reason that international intervention was rejected was that the conflict was sold as an intractable ethnically-inspired confrontation. The Serbs sold the very notion of constructive mediation as unworkable. Furthermore, if the external powers did intervene, they would be made to bleed for their mistake.

Drawing on the partisan history of the JNA, the Serbs were able to convey the impression that they were willing to wear down any intervening force in a long guerrilla war. The ethnic and therefore seemingly irrational warfare that resulted was one of the reasons that the new war thinkers heralded these wars as post-Clausewitzian. As the Military historian John Keegan put it, 'by their nature these wars defy efforts at mediation from outside … they are apolitical'.[48] This fit the Serb war aims perfectly; as long as the international community believed that the wars were intractable and that the belligerents would put up a fight, the chances of them intervening robustly were slim. This was a political tactic to avoid international action. As we know already, in reality the Serb forces suffered from acute problems in terms of supplies and in morale. Nevertheless, they successfully managed to maintain the illusion that they were a guerrilla force capable of defeating the great powers should they intervene militarily.[49] Though the Serbs successfully presented their forces as new-age partisans to sections of the international community, the consequences of their tactics presaged firmer action with a more robust policy being followed from 1994 to 1995, led by the US.

As noted above, the US had sat on the sidelines throughout much of these conflicts. When it was involved, it consistently opted for the use of airpower and thus clashed with the British and French position. Like the British, the French and the Dutch – the main troop-contributing states in UNPROFOR – worried that escalation would increase the likelihood that UN personnel would be targeted by the combatants; each had been subjected to the humiliation of having personnel held hostage. Conversely, the US did not have any troops in theatre with which to complicate their attraction to using air-power. The US shunned the notion of greater involvement in the conflict until 1994, and was never sold on the idea of direct military intervention other than air-power. Nevertheless, a more robust and far-reaching US policy resulted from several interlaced factors. American involvement gradually strengthened as the war became protracted and as it became clear that European attempts at ending the conflict were proving futile. The failure of the Europeans to bring about a suitable conclusion to the war through UN channels resulted in renewed calls for stronger NATO participation. Moreover, if the US continued to sit on the fence it risked losing its status as the leader

---

48   John Keegan, *A History of Warfare* (London: Pimlico, 1994), 58.

49   See: Norman Cigar, 'The Serb Guerrilla Option and the Yugoslav Wars: Assessing the Threat and Crafting Foreign Policy', *Journal of Slavic Military Studies*, 17(3) (2004), 485–562.

of NATO, derailing the alliance's outreach to former Soviet states, which were in the process of being brought closer into the Western sphere of influence. The failure to act may have even put NATO's very future at stake.[50] Commenting at the time, the then Secretary General of NATO Manfred Woerner tried to provoke a greater NATO role. He noted that 'We all wish that diplomatic means alone would succeed. But diplomacy needs to be backed up with determination to use force if this is to be credible ... In short: you need NATO. The United Nations are overstretched and underfunded'.[51] Although Woerner died before NATO acted decisively in Operation Deliberate Force in August 1995, there was increasing pressure for NATO to take a key role. There was also a feeling that the US needed to exert its authority if NATO was to retain its legitimacy. For the US and NATO to consolidate their Cold War victory they needed to shore up their credibility and legitimacy. Leaving BiH to implode completely was not an option. This was the view of Richard Holbrooke, the Clinton Administration's special envoy to Bosnia-Herzegovina. In a memorandum sent by Holbrooke prior to his position as envoy, he warned that:

> Bosnia will be the key test of American policy in Europe. We must therefore succeed in whatever we attempt. The Administration cannot afford to begin with either an international disaster or a quagmire. Despite the difficulties and risks involved, I believe that inaction or a continuation of the Bush policies in Bosnia by the Clinton Administration is the least desirable course. Continued inaction carries long-term risks which could be disruptive to US-European relations, weaken NATO, increase tensions between Greece and Turkey, and cause havoc with Moscow ...[52]

Holbrooke would later remark, 'It was not an overstatement to say that America's post-World War II security role in Europe was at stake'.[53] Actually, exemplifying the very nature of chance and uncertainty, especially the notion that small decisions can have big effects, the US had agreed to provide the bulk – around 20,000 troops – of personnel to oversee the UN withdrawal should the need arise. Paradoxically, although the UNPROFOR risked collapsing in the spring of 1995, largely because it lacked clear policy or US military muscle, UN withdrawal would trigger the deployment of US ground troops to Bosnia.[54] The prospect of US troops being deployed as part of a withdrawal and thus signifying international failure was inconceivable if the US wanted to retain its international legitimacy. As Holbrooke

50   Ryan Hendrickson, 'Leadership at NATO: Secretary General Manfred Woerner and the Crisis in Bosnia', *The Journal of Strategic Studies*, 27(3) (September 2004), 515.

51   Cited in Ryan Hendrickson, 'Leadership at NATO', 515.

52   Holbrooke, *To End a War*, 50.

53   Holbrooke, *To End a War*, 67.

54   Plan OpPlan 40–104 had the authority to bypass the President, who was only informed of it by Holbrooke.

recounts, 'it was a terrible set of choices, but there was no way Washington could avoid involvement much longer'.[55] Although it took until the summer of 1995, a combination of US political power and NATO firepower put the Serbs under mounting pressure.[56]

NATO's military intervention was not the primary reason for the reversal of fortune for the VRS in August and September 1995. That is more directly accredited to the coordinated offensives of the re-established Croat-Muslim alliance. Nevertheless, the two are intimately linked. The Croat and Muslim offensive was not only the product of Croat and Muslim bellicosity towards the Serbs; the very fact that there was an alliance at all was attributable to the prescience of the Washington Agreement, which successfully brought about a Croat-Muslim rapprochement in 1994. With the backing of the West, the Croat-Muslim offensives in 1995 were well organized and executed. According to one UN observer, operation STORM, the military offensive which drove the Army of the Serbian Krajina (OS-RSK) and the Serb population out of the Krajina in 1995, illustrated the closer ties between Croatia and the West: 'Whoever wrote that plan of attack could have gone to any NATO staff college in North America or Western Europe and scored an A-Plus'.[57]

Although the Bosnian Muslims remained the least able to profit from the collective international malaise, Croatia did profit. As pointed out in the previous chapter, the Vance-Owen Peace Plan effectively handed over Western-Herzegovina, and it was viewed, and used, by the Croatian forces in BiH and their political masters in Zagreb as a *fait accompli*; the de facto annexation of Bosnian Croat territory. Although Zagreb fiercely defended the territorial integrity of Serb-held regions of Croatia, Tudjman's ambitions towards Bosnia-Herzegovina and the half-hearted attempts by the international community to find a solution to the conflict all contributed to Croatia's territorial aggrandizement at the expense of the Bosnian Muslims. In terms of the VOPP agreement, the conflation of hostility and chance provided the Croats with the opportunity to pursue a more aggressive policy at the expense of the Bosniaks, condemning the idea of a unified multi-ethnic Yugoslavia in the process. Brendan Simms directly links this opportunism to the failure of the international community to halt the fighting early in the war. One of the fiercest critics of the lack of external action, Simms notes that,

---

55   Holbrooke, *To End A War*, 65–6.

56   Important works on NATO transformation include: Rachel A. Epstein, 'NATO Enlargement and the Spread of Democracy: Evidence and Expectations', *Security Studies* (2004–5); Christopher Coker, 'Globalisation and Security in the Twenty-first Century: NATO and the Management of Risk', *Adelphi Paper*, Issue, 345 (Oxford: Oxford University Press, 2002). Williams, Michael J., *NATO, Security and Risk Management. From Kosovo to Kandahar* (Abingdon: Routledge, 2009).

57   Colonel Leslie, cited in Silber and Little, *The Death of Yugoslavia*, 357.

Once it became clear that no western military intervention would materialise, the Bosnian Croats felt emboldened to embark on their own separatist project. Indeed, they were almost forced to do so by the influx of Muslim victims of unchecked Serb aggression.[58]

Although Zagreb's policy regarding Bosnia mirrored that of Serbia and its cohorts in Bosnia, Zagreb was more in tune with the twists and turns of war as it unfolded. Whether this was down to prescience or mere happenstance is unclear. However, the Croats were more amenable to the motives and threats of the external powers. Zagreb appears to have been quick to take opportunities when they arose. By the start of the Dayton negotiations, Holbrooke admitted that Tudjman and Croatia had become the key factors in finding a solution. Reflecting on an earlier example, in a move which would have far-reaching consequences for the outcome of the war, the US inspired Washington Agreement in 1994 brought an end to the Croat-Muslim war in Western and Central Bosnia. Although the combatants remained suspicious of each other, the end of hostilities was quickly followed by the re-establishment of the Croatia-Bosnian Alliance, which in the late summer of 1995 proved so injurious to the idea of a 'Greater Serbian' State. From humble and unlikely beginnings, this alliance forced the Bosnian Serbs back under the control of Milošević's Serbia, and ultimately, to a negotiated peace at Dayton.

The Croats had much to lose if they did not reach agreement over rapprochement with the Bosniaks. The US position was clear; it reminded Zagreb that it was committed to Croatian territorial integrity, but would not back Zagreb if it continued to annex territory legally belonging to the Bosnian government. If they did come to an agreement, Tudjman would have much-needed US support regarding the autonomous Serb regions in Croatia. Croatia would also benefit from closer political ties, essential economic aid, and the opportunity to take its place as a European state. If Tudjman complied, the rewards were economic and military integration into Western political and security organizations. If Tudjman refused, like Serbia, Croatia would become an international pariah.[59]

Assessing the Bosnian Muslim situation, it is clear that there was a more limited range of options available. Even in terms of the conflict with the HVO, although the ARBIH had a large quantitative advantage over the Croat forces, the HVO was backed by the increasingly powerful Croatian Army and had access to a better selection of arms and supplies. If the ARBiH continued to fight on two fronts, against the Croats on one hand and the Serbs on the other, then it risked being strangled by the forces ranged against it. Although international rhetoric frequently offered solace to the Bosnian Muslim position, increasingly berating the Serbs in particular for their aggressive behaviour, the Bosnians had to rely on their own strength to survive the worst days of the war. Thus, they too must be given some credit for seeing the potential of the Washington Agreement.

58 Simms, *Unfinest Hour*, 33.
59 Silber, *The Death of Yugoslavia*, 322.

The extent to which this was in fact the case was not clear when the two sides agreed to broker a ceasefire and re-constitute the Croat-Muslim alliance. The Croat-Muslim war had been particularly dirty and displayed ethnic-cleansing and attacks against cultural and religious symbols. Most poignantly, the wanton destruction of the medieval Turkish bridge at Mostar by Croatian artillery conveyed the level of suspicion and hostility which was sustaining the conflict. It is important to add that the significance of the Washington Agreement was not immediately clear, and although the Croat and Bosnian forces had tacitly agreed to suspend hostilities, a great amount of anger persisted on both sides. In fact, as Hoare has explained, rather than bringing the two communities together under a unified command structure which could then attack the Serbs collectively, the alliance was managed by a loose organization – the newfound rapprochement between Muslims and Croats was surface deep. Instead of bringing the sides together, the ARBiH increasingly recoiled from its original mandate to be the army of a unified Bosnia, and instead evolved into a party army of the SDA. Although outwardly Muslims and Croats shared political control, the ARBiH supported the SDA as a bulwark against anti-Muslim machinations.[60] Feelings were running extremely high and the wounds of the Croat-Muslim conflict are still very much evident today. Despite this, in order to defeat the Serbs and thus rescue hope of a Bosnian state, the Bosnian government had to put aside its distaste for the Croats and sign up to the Washington plan as engineered by Charles Redman and Peter Galbraith.

Chance, and the Problem with Probabilities

Chance and uncertainty exist within the matrix of real war and it is notoriously difficult to accurately mark exactly when and how this tendency impacts on war when it is ongoing. This is the problem posed to the commander and the statesman, and it is why Clausewitz believed that the traits of a genius were required to overcome the complexities which war naturally displays. Clausewitz reminds readers, 'The conduct of war branches out in almost all directions and has no definite limits'.[61] This is why it is so difficult to accurately delineate an easy path to victory. Neutral factors such as chance complicate the existing complexities highlighted by Clausewitz through his inclusion of hostility – passion, hatred, and enmity. In fact, chance finds its form in that interaction. The omnipresent, invasive characteristics of chance as it evolves through a constant reciprocal interaction are hard to distinguish. It is the resulting uncertainty that is the key to understanding the intrinsic, symbiotic nexus which exists between escalation – the interplay of 'opposites' – and the pivotal position which chance enjoys within the Trinity. All war is pervaded by uncertainty and therefore the actions of all actors and

---

60   Marko Attila Hoare, *How Bosnia Armed* (London: Saqi Books, 2004), 107–8.
61   Clausewitz, *On War*, 154.

combatants are restricted, buoyed, and directed by chance encounters. To quote Clausewitz once more, belligerents make decisions in 'a fog of greater or lesser uncertainty' and these decisions can have profound, if unlikely, impacts on the outcome of particular wars. The point Clausewitz was making is that once war is underway, the complex interrelationship between cause-and-effect produces an infinite number of consequences which must be judged and overcome, as all the while these factors continue to produce new problems, driving the war down unexpected routes.

In summation, the purpose of this chapter has been to draw out the complexity of Clausewitz's conception of chance, and explore its relationship as part of the Trinitarian nature of war. By placing the assessment of chance against the background of wider external diplomacy and intervention in the wars of former Yugoslavia, it has been illustrated that uncertainty provided disadvantages and opportunities for each of the combatant groups. In addition, by evaluating the role of chance and uncertainty as part of the interactive reality of conflict, it has been possible to highlight the array of intersecting lines which feed into and develop the nature of war. It is axiomatic that the involvement of the international community contributed to the complexity of the conflict and it undoubtedly produced opportunities for the belligerents on the ground. These opportunities, of course, were not always clear and thus there was a danger that the tactics used to make the most of strategic opportunities could rebound at some later date. This is exactly what happened to the Serbs. It was a problem with which Clausewitz was familiar. He cautions against rash use of force delinked from war as a 'whole': 'We must evaluate the political sympathies of the other states and the effect war may have on them.'[62] The problem in doing this is made so much more difficult as actors and motives interact and change. Clausewitz continues: 'To assess these things in all their ramifications and diversity is plainly a colossal task … Bonaparte was quite right when he said that that Newton himself would quail before the algebraic problems it could pose.' It was this capricious nature that brought Clausewitz to surmise that 'genius' – the natural ability of the commander to react to chance and overcome complexity by acting intuitively – was vital if strategy was ever to translate into victory. Indeed, it was this type of fluidity that provoked social scientist Richard K. Betts to ask whether strategy was actually an 'illusion'.[63]

The Serbs took advantage of the initial inability and unwillingness of the international community to engage in the conflict militarily. In the early stages of the war this offered an opportunity and the Serbs maximized their military advantage. Although it was evident that the external powers would eventually oversee the war's termination, it was likely that the lacklustre efforts of the international community would take the situation on the ground as a *fait accompli*. The Serbs pushed home their advantage. The problem with this approach was

---

62   Clausewitz, *On War*, 586.

63   Richard K. Betts, 'Is Strategy an Illusion?', *International Security*, 25(2) (Fall, 2000), 5–55.

that they failed to constantly reassess the international community, its resolve and its growing irritation with themselves. Although their intransigence won rewards initially, it backfired in the long term and helped to gradually turn international opinion against them. As Serbia become isolated, it became more reckless, reinforcing international anger.

Although chance can offer rich rewards, the uncertainty with which it is entwined also produces dangers which have the potential to wreck previously hard-earned gains. It is a problem that Clausewitz knew well, and is explicit in his conceptualization of the culminating point of success. The commander – or statesmen – must calculate what the best option will be, and then make the correct decision. Unfortunately for the Serbs, they failed to perceive the often subtle but changing circumstances in the international community, or equate these changes to the course of the war on the ground. While they sought to maximize their advantage, their actions inadvertently isolated them and pushed the external powers to embrace their enemies. They were caught out by a recurring pitfall – lack of verifiable intelligence. As Clausewitz reflects, the leader must:

> Guess whether the first shock of battle will steal the enemy's resolve and stiffen his resistance, or whether, like a Bologna flask, it will shatter as soon as its surface is scratched; guess the extent of debilitation and paralysis that the drying up of particular sources of supply and the severing of certain lines of communication will cause the enemy; guess whether the burning pain of the injury he has been dealt will make the enemy collapse or, like a wounded bull, arouse his rage; guess whether the other powers will be frightened or indignant, and whether and which political alliances will be dissolved or formed. When we realize that he must hit upon all this and much more by means of his discreet judgement, as a marksman hits his target, we must admit that such an accomplishment of the human mind is no small achievement. Thousands of wrong turns running in all directions tempt his perception; and if the range, confusion and complexity of the issues are not enough to overwhelm him the dangers and responsibilities may.[64]

The VRS passed the culminating point of their success. Although it may not have been particularly apparent at the time, the Washington Agreement and the resulting rapprochement between the formerly warring Croat and Muslim factions eventually brought the Serbs to the negotiating table. This is only one of many examples, but it serves to highlight how the course of a particular war can change through seemingly small ineffectual measures. The success of the Bosnian-Croat Federation hinged on the ceasefire between the two sides, and would not have been possible if it had not been for US diplomatic efforts. In the same vein, this rapprochement would not have had the successful results that it did if the US and NATO had not become more centrally involved.

---

64   Clausewitz, *On War*, 692–3.

Moreover, that both the US and NATO did become centrally involved in the conflict had as much to do with US standing and NATO enlargement, of finding a new role in the post-Cold War security environment, as it did with alleviating the suffering in Bosnia-Herzegovina. The events and outcomes are all intertwined. The nature and course of a particular conflict is influenced by a dazzling array of factors, not always immediately obvious, but which are produced by, and further produce, the complexity of conflict – in Clausewitz's terms, a Trinity.

Exposing the unpredictable course of war once again, it is useful to ponder further on how seemingly small events can go on to produce profound results. Although the Washington Agreement ultimately proved a successful mechanism for balancing the Serbs, it was not until the US felt compelled to take a leading role as a result of international pressure that real progress was made. It was the shuttle diplomacy, led by special envoy Richard Holbrooke, which eventually brought the competing sides to agree to the Dayton peace agreement. This was made considerably easier because of battlefield events. Although US diplomacy resurrected the peace process, this process was considerably propelled by the new-found operational coordination between the HVO and ARBiH during 1995. The prospects for a lasting peace were also bolstered, ironically, by the fact that territorial ethnic cleansing had also resulted in acceptable gains for the Serbs, who were now willing to negotiate the termination of the war. Although the seizure of Srebrenica and Žepa had eventually instigated a firmer international role, the fact that the Serbs had what they wanted made it easier to negotiate with them. Despite the label 'safe area', it was widely accepted by the external powers and Bosnian government that the enclaves were ultimately indefensible.[65] For the Serbs, the enclaves posed a constant threat to the security of Republika Srpska. Despite international censure, the seizure of the safe areas actually made the prospect of a negotiated settlement a more realistic option.[66] It is evident that while changes at an international level brought unexpected results – not least the cessation of hostilities, chance and uncertainty pervaded the decision-making process of the external powers at every juncture. The everyday interactions of international relations, of relationships between states, international organizations, and indeed individuals, all contributed to a feeling of malaise which fed into the war; these were unexpected and often unnoticed features, but they influenced the nature of the conflict nonetheless. As ever, the events that brought the external powers and

---

65   For example, the Bosnian government withdrew the military leadership from the enclave before the Serb offensive. It has been suggested that the Bosnian government gave up Srebrenica so as to engender international condemnation of the Serbs.

66   See: Nicholas Sambanis, 'Partition as a Solution to Ethnic War: An Empirical Critique of the Theoretical Literature', *World Politics*, 52(4) (July, 2000), 437–83; Other works include: Chaim Kaufmann, 'Possible and Impossible Solutions to Ethnic Civil Wars', *International Security*, 20(4) (Spring, 1996), 136–75; Thomas Chapman and Philip G. Roeder, 'Partition as a Solution to Wars of Nationalism: The Importance of Institutions', *American Political Science Review*, 101(4) (2007), 677–91.

combatants to that end are inextricably intertwined with the perennial interaction of the belligerents and the forces of hostility, chance and policy.

In Clausewitz's Trinity, chance exemplifies the unforeseen within the constantly oscillating nature of war. It is a natural by-product of war itself and the commander and political leaders require special acuity if they are to overcome its mercurial tendency to pervert the very best of strategies. The manner and implications of this element do not, as Clausewitz notes, 'yield to academic wisdom.[67] They cannot be classified or counted. They have to be seen or felt'. Determining whether the convergence of hostility and chance pervades and shapes policy, or merely 'jostles' it about, is the subject of the following chapter.

---

67   Clausewitz, *On War*, 216.

Chapter 5
Policy

The Clausewitzian Trinity: War as a Continuation of Policy?

The previous two chapters have examined the ways in which hostility and chance conflate and influence conflict. The present chapter analyses the influence of these factors on policy. In what way does the inherent interaction of the Trinity change policy? Is policy really 'a priori' equal with the other elements within the Trinity? By drawing on the discussion in the previous chapters, the present determines the place of 'policy' in the Trinitarian construct and provides the basis from which an intelligent analysis of the strengths and weaknesses of the concept can be made.

Although the interactivity at the heart of the Trinity is genuinely perceptive, does Clausewitz's concept hold true when assessed against the complexity of real war? This is an important point of debate and recent analysis by leading Clausewitzian scholars appears to share the view that the Trinity is comprised of three equal elements – hostility, chance, and policy. In fact Clausewitz tells us that:

> These three tendencies are like three different codes of law, deep rooted in their subject and yet variable in their relationship to one another. A theory that ignores any one of them or seeks to fix an arbitrary relationship between them would conflict with reality to such an extent that for this reason alone it would be totally useless.[1]

He continues: 'Our task therefore is to develop a theory that maintains a balance between these three tendencies, like an object suspended between three magnets'.[2] It would seem that the key to understanding the complex nature of war is by understanding the interconnectivity of the competing elements of the Trinity: they are equal. As noted in the introductory chapters, however, this clashes with Clausewitz's insistence that war is an act of policy. It was this pronouncement which has engendered such enduring interest in his ideas. Indeed, his decision to revise On War in 1827 was primarily because he realized that war was a continuation of policy. The political component was the key to understanding the purpose and nature of war – war's instrumentality derives from its indelible nexus with policy. However, if policy is caught in a perpetual struggle with chance and hostility, then does war not lose its instrumentality? This marks a clear disjuncture with the idea that war is instrumental, a continuation of policy. It was this idea which invested

---

1   Clausewitz, *On War*, 101.
2   Clausewitz, *On War*, 101.

Clausewitzian theory with its universal value, and which seems to resonate so powerfully throughout the history of war. It is the reason why Clausewitz's tome has retained relevance despite constant changes to the way war is fought. Did Clausewitz really intend his Trinity to be comprehended in this way?

As the final part of the case study, the present chapter is designed to follow on from the discussion in the previous two chapters and explores the role of policy in juxtaposition to hostility and chance. In the chapters dealing specifically with these elements the purpose was to examine the cause-and-effect relationship of each phenomenon and to draw out examples, assessing the ways in which they can manifest themselves in war. This chapter draws on that discussion in an attempt to gauge the extent to which policy is influenced by the non-linear elements of hostility and chance. Like the preceding chapters, it does not provide a historical exploration of the evolution of policy within the competing groups in the wars in question. Neither does it attempt to be a historical essay listing the daily twists and turns of political life. It is not a recreation or re-enactment of the political decision-making process. Divided into three primary sections, the chapter identifies the original purpose of each of the main belligerents at the start of the conflict and assesses the impact of the Trinity's reciprocity on these aims as the war progresses. Comparison of the combatants' original motives with the eventual end state reveals the Trinity's influence.

Serbian Policy, and the Trinity

What were Serbia's war aims at the start of the conflict in 1991? Did they stay constant or did they shift as the war evolved? Although it was never officially declared, there is wide consensus that in 1991 the Serbian war aim was to create a 'Greater Serbia' which would house all Serbs in one single state. This was a long-term Serbian aspiration that had been partially accomplished by Serbia's inclusion within the Kingdom of Serbs, Croats, and Slovenes in 1919 and it was at risk if Yugoslavia disintegrated. The large Serbian minorities, particularly in Croatia and Bosnia-Herzegovina, would be cut-off from Serbia proper should dissolution occur and this was something that the Serbian government would not acquiesce to. In the starkest terms, Milošević's regime chose war as a means of determining the favourable position of state boundaries following the dissolution process. Nevertheless, Serbian policy went through several stages and altered as the fluid character of the conflict shaped and twisted the original motive.

As the competing components of the Trinity met and conflated, the initial Serb war aim evolved to meet the changing reality of the conflict. As James Gow has correctly observed, the Serbs had a project which reacted to events rather than a single coherent plan.[3] This reflected both the confusion surrounding the early

---

3   James Gow, *The Serbian Project and its Adversaries* (London: Hurst & Co., 2003), 12.

passage of the war, and a pre-existing weakness in Serbia's political position. For instance, even as the war began the JNA and Serbia had not completely cemented their relationship, only converging properly as the vicissitudes of real war shaped the political decision-making process. The army and Serb irregulars would fall back on Serb majority areas with the eventual purpose of subsuming these territories into Serbia. However, when Croatian resistance moved from static and disorganized defence into a more robust fighting force capable of offensive action of its own, Milošević changed tack and allowed UN peacekeepers to take on the task of policing the new Serb-Croat frontier. A combination of interlinked factors ranging from dogged Croatian resistance to poor Serbian mobilization rates and international pressure would prevent Serbia from attaining its territorial aspirations in Croatia. The Croatian Serbs would live in a Serbian state, but that would sit outside Croatia's republican borders prior to the war. This ability to change direction and change in accordance with events would underpin the entire Serbian approach to the war.

This mirrors the conflict in Bosnia-Herzegovina. As the war moved over the border, Serbian policy was to annex territory which would then form the basis of an enlarged Serbian state. As highlighted in Chapter 4, Serbia's military campaign had early success which resulted in rich rewards – by the close of 1992 Serb forces held around 70 per cent of BiH territory. However, the Serbian war aims were never fully met. The very element of power which had first enabled the Serbs to annex tracts of Bosnia – military force – eventually worked against them. To restate an earlier quote. in the words of Clausewitz, 'the will is directed at an animate object that reacts'. The original Serbian purpose and early military successes did not go unchecked and a combination of factors intersected to work against Serb interests. Not least, ethnic cleansing and barbarism provoked local resistance and international outrage. This combination of opposition and the outpouring of international condemnation, which ensued chiefly from Serb tactics and intransigence, resulted in a formidable grouping of enemies, eventually paving the way for the unlikely coalition of the HVO, ARBiH, and NATO to roll back Serbian forces from their 1992 position.

From an initial policy which intended to carve off 60 per cent of Bosnian territory for incorporation into an enlarged Serb state, the war produced a schism in the Serb nation, and a combination of international pressure and military reality resulted in the unravelling of Serb policy.[4] It would be reoriented to fit the changing balance of power in former Yugoslavia. However, rather than bringing all Serbs into one state, the war ended with the expulsion of the Serbs from Croatia and a pariah statelet – Republika Srpska – in Bosnia which was cut-off from Serbia proper. In Belgrade, Milošević's earlier nationalism gave way to a reframed socialism; in the process castigating his proxies for international outrages and

---

4   Although the Serbian vision was never stated explicitly, Stipe Mesić recounts that during one discussion with Borislav Jović, Jović indicated that the Serbs were preparing to annex 60 per cent of BiH.

crimes against humanity. It was a very different policy from the one which started the war, standing in stark contrast to the infamous proclamation to the Serbs of Kosovo that 'no one will dare to beat you'.

These changes in policy direction are directly attributable to the consequential effects of the maelstrom of conflict in Croatia and Bosnia-Herzegovina. As the competing tendencies conflated, the original purpose of the war evolved. The interaction of opposites, an enemy which reacts to aggression, to action and counter-reaction, produces the reciprocity exemplified by hostility and chance. As the previous chapters explain, these tendencies collide, conflating and infusing the original motive with opportunities and complications alike. The vortices of war explicated by the Trinity are instructive. However, it is vital to identify that as war is born by policy, so too is policy shaped and changed by war. As was the case with Serbia, a process of continuous feedback ensures that policy can adapt to meet the evolution of a particular conflict. It is clear that the notion that war is a purely rational means-ends activity is undermined by the interaction of 'real' war. However, because policy is apt to evolve as it reacts to war's changing nature it preserves some degree of instrumentality. It is up to policy-makers and military commanders to account for the evolution of the conflict and to realign strategy to reflect this, which may not be easy. Yet the complex web of relationships and feedback between the nature of war, policy, and strategy, ensures that, in some shape or form, policy is always guiding war's own evolving nature; as it does war's character also.

Although Serbia had legitimate grievances regarding the constitutional alterations of the Tito era, the rise of Milošević and the rise of Serb nationalism which accompanied it proved particularly damaging to the long-term success of the Yugoslav idea. Advocating a centralist reorientation of the Yugoslav state, the Serbian national programme produced a deleterious national backlash in the other federal republics, putting them on a collision course towards eventual war.

By its very nature, the political fragmentation that presaged the beginning of the war was explosive and confused. It was not always clear what was happening, or who was fighting who. Although we can identify an underlying Serbian national programme – with the aim of creating a Greater Serb state, this also mirrored the JNA ambition to prevent the disintegration of Yugoslavia by overthrowing the recalcitrant state leaderships in the newly independent republics of Slovenia and Croatia. The army remained quasi-independent from Milošević, and only converged with Serbian policy as the crisis turned into all-out war. The Serbian policy of territorial aggrandizement mirrored early JNA policy and brought the army further into the Serbian strategic ambit.[5] From the outset of hostilities in

---

5   Stipe Mesić, *Danas*, June 24 and 25, 2000; cited in Lenard J. Cohen, *Serpent in the Bosom: The Rise and Fall of Slobodan Milošević* (Boulder: Westview Press, 2002), 192. In a similar vein, James Gow has argued that one function of the paramilitary groups used in Croatia and BiH was to force the JNA to follow Serbian policy – Arkan's Tigers and Šešelji's Chetniks were put into action in the knowledge that the JNA would support

1991 the firepower of the regular army provided the Bosnian Serbs with a huge military advantage which would provide 'the basis of their ability to conduct military activity' throughout the entire war.[6]

As highlighted in Chapter 4, although the causes of the conflict are complex, one major point of contention between the republics, and especially between Serbia, Croatia and BiH, revolved around whether the Serb population in Croatia and Bosnia-Herzegovina should have the right to secede in order to join Serbia. As a consequence of this, Serbia's early policy was aimed at attaining the territories which held substantial Serb populations – 12 per cent and 31.2 per cent of Croatia and BiH respectively. A quick annexation of Croatian territory was critical if Milošević was to fulfil his policy objective. Although the initial justification for JNA involvement was simply to protect the peoples of Yugoslavia, its real aims were to safeguard Yugoslavia's territorial integrity. The purpose of its operational involvement therefore had a practical element: Serbian irregulars and the JNA would carve off as much territory as possible. As Norman Cigar argues, 'war was a conscious policy decision which was necessary if he (Milošević) was to change the existing republican borders in order to achieve his stated goal of 'all Serbs in one State'.[7]

Despite initial successes, the overwhelming firepower at the disposal of the Serbs did not result in outright victory and the early policy of annexing large tracts of Croatian territory was partially prevented. Consequently, by January 1992 Serbian policy had changed from one advocating the destruction of Croatia to one which favoured holding on to existing gains. Stiffer than expected Croatian resistance, especially around the strategically important centres of Vukovar and the Slavonian hinterland, as well as mounting international condemnation of Serbia's policy, had important effects and are directly linked to Milošević's decision to seek a UN-negotiated settlement to the war in Croatia. Serbia and Croatia signed the VOPP in Sarajevo on 2 January 1992 bringing about an end of the first phase of major fighting in the wars of Yugoslav dissolution. Early success in flushing out the Croat population of Krajina was slowed down in the later part of 1991, and by September the Croats had recovered from a shaky start and formed a cohesive frontline stretching from Dubrovnik to Vukovar. As initial successes turned into setbacks, Milošević altered the Serbian war aims.

At this juncture the conflict demonstrates the kind of escalatory features which Clausewitz thought must necessarily require greater amounts of effort from each opponent. Serbia's aggressive stance and its active support for the Serbian irregulars in the Krajina produced an equally forceful riposte. This is exactly the

---

them when Croatian forces retaliated, further bringing them under Serbian influence. Gow, *The Serbian Project*, 79–89.

6   Norman Cigar, 'Serb War Effort and Termination', in Branka Magaš and Ivo Žanić (eds), *The War in Croatia and Bosnia-Herzegovina, 1991–1995* (London: Frank Cass, 2001), 211.

7   Norman Cigar, 'Serb War Effort and Termination', 205.

sort of interaction that, until the revision of his work in 1827, Clausewitz believed should precipitate escalation. The side which was willing to escalate fastest and which had the most power should win. This is the message he advances in the opening chapter of his work when he discusses the *interactions to the extreme*. As he explains: 'To introduce the principle of moderation into the theory of war itself would always lead to logical absurdity.'[8]

However, the reason Clausewitz began his revision of *On War* was because he was cognizant that escalation was prohibited by a complicated mixture of politics and friction. The escalation of the war in the Krajina only went so far, and it did so because policy changed to meet the changing, broader, political environment in which the war was fought. Furthermore, this change can best be understood by a deeper comprehension of the dynamism at the heart of war's nature. Far from being held captive by the competing components in the formula, there was continual feedback relaying information back and forth between Belgrade and the theatre of operations. An updated policy re-engages with the ongoing war, itself affecting and changing the nature and character of the conflict – that process is continuous. The level of Croatian resistance forced Milošević to update and refine his aims in the light of the realities on the ground. Rather than escalate further and risk a protracted conflict which would sap Serbia's political and material resolve, Milošević sought to end the war on the most favourable terms possible; exactly the point made by Jovan Jović, when he explains the rationale behind Serbia's decision to utilize the involvement and intervention of the international community in the shape of a UN protection force. Serb forces used the UN as a police force which would prohibit Croatian incursions, closing off a potentially wasteful escalation of the war and allowing the Serbs to concentrate on the bigger prize – Bosnia-Herzegovina.

Although the war in Croatia is filled with examples of hostility and battle, the fight for Vukovar serves as an emotive symbol of resistance, providing the empirical proof that war on paper is something quite different to real war. Although the JNA and Serb irregulars had a significant qualitative advantage, they suffered from debilitating morale problems and had staffing shortages in key areas – a situation exacerbated by desertion and draft-dodging on a massive scale. As the Serbian Ministry of Defence put it, during 1991 the draft-dodging was to such an extent that 'There was a danger that Serbia would have to defend itself in Zemun'.[9] This stands in sharp contrast to the mood in Croatia, and it is evident that the Serb's qualitative military advantage was reduced and less effective in the face of stubborn Croatian resistance. Under these conditions it was deemed that Serb forces would be unable to break down the Croat defence at an acceptable cost. Of course, the decision to end the war in Croatia reflected the deepening crisis in Bosnia. Nevertheless, it is hard to ignore the direct link between Croatian

---

8   Clausewitz, *On War*, 84–5.
9   Norman Cigar, 'Serb War Effort and Termination', 206. The town of Zemun lies on the outskirts of Belgrade.

resistance – itself a direct result of war's natural propensity to engender hostility – and Milošević's willingness to modernize his objectives. Not only does this highlight the importance of the interplay between the competing elements of the Trinity, it is a clear example of interaction and reciprocity shaping, and indeed precipitating, a modification of purpose.

Serbian Policy in Bosnia-Herzegovina

Although many new war theorists use the war in Bosnia-Herzegovina as an example of the changing nature of war, this is partly due to the fact that they look at one phase of a wider conflict, treating it as an individual war. Yet this dislocation of the conflict from the wider issue of Yugoslav dissolution, especially the initial phases of fighting in the neighbouring republics, fails to properly understand the true complexity of the war. The conflict was part of a wider conflagration and as such the wars stretching from Slovenia in 1991 to Kosovo in 1999 are indelibly linked. As in Croatia, Serbian policy in Bosnia was to annex as much territory as possible. Again mirroring the purpose of the conflict in Croatia, this would then form the basis of an expanded 'Greater Serbia'.

At the beginning of the hostilities in Bosnia, the Serbs utilized their superior military capability, forcibly clearing large swaths of territory populated by non-Serbs. While the military capability of the regular army provided the firepower, Serbian paramilitary groups such Arkan's Tigers and Šešelj's Chetniks were instrumental in fomenting hostility by engendering fear and hatred in non-Serb communities. In towns such as Bijeljina and Zvornik the existing population was uprooted – the lucky ones were forced to leave.[10] Through this mixture of intimidation and murder, Serbian forces were able to utilize their qualitative advantage and occupy at one point as much as seventy per cent of the territory of BiH. Although the rationale for the early successes was strategic, going a long way to securing the future of Republika Srpska, the manner of these victories acted as the prelude to ethnic cleansing, which became one of the central characteristics of the conflict.

Despite these early 'victories', Bosnian Serb success proved ephemeral and they failed to win a knockout blow against the Bosnian Muslim or Croatian forces. Furthermore, as the war became protracted its original purpose became stretched and then changed, in accordance with the consequences of interaction. Just as the plan had altered in Croatia, if we contrast the initial policy against the final outcome we get the clearest affirmation that policy changed and adapted as the different forces of war conflated. In fact, revision of policy continued almost constantly. During Milošević's summit with Tudjman prior to the outbreak of war

---

10 In Zvornik, some 2,000 people are still unaccounted for. A further 47,000 were expelled from the area, becoming refugees and adding to and exacerbating the deteriorating societal tensions already caused by the war.

in Croatia, the two leaders had agreed on partitioning the Bosnian republic. The Serbs, being the most powerful and having a far bigger ethnic population, would naturally take the biggest share. At the end of the conflict the Serbs had to settle for significantly less than they had originally forecast – 49 per cent. Until the late summer of 1995, they had held around two-thirds of the territory of BiH.

This ongoing re-correlation of policy can be viewed by examining Belgrade's relationship with the Bosnian Serb leadership. Although it was the patronage of Milošević which had endowed them with political and military power, the war opened up political fissures which complicated the original Serbian motive as espoused by Milošević. On the one hand, the opening and deepening of this schism represents irrationality, and at first it seems to suggest a propensity to quarrel.[11] It certainly complicates the analysis of Serbian policy as a result. Instead of one coherent Serbian policy, there were two competing 'projects' working against each other. On the other hand, however, although this schism could be interpreted as an exemplar of the conflict's underlying irrationality, perpetuated by petty squabbles and rapacity, it also masks a more complex and intriguing explication of Clausewitz's assertion that reciprocity breeds complexity and unpredictability. Clausewitz may have concurred with the new war writers: a great many of war's traits are intrinsically irrational. Nonetheless, irrational or not, the growing fissures between Belgrade and Pale (the Bosnian Serb wartime capital) were themselves the product of war's non-linear nature producing a strategic milieu at odds with Belgrade's wishes; a fascinating insight which captures how war's inherent but tumultuous forces have the potential to tear up the rule book. The schism represents much more than a propensity to quarrel. It reflects the complexity of real war, and especially the difficulty in corralling passions and egos once they have been released and shaped by fighting.

Milošević's campaign to grab territory in Bosnia-Herzegovina began prior to Bosnia's referendum. It was Belgrade that shaped Bosnian Serb politics in the years and months prior to the war and it was Belgrade that armed the Bosnian Serbs in readiness should war break out. Immediately before the opening salvos of the Bosnian war, Milošević and Jović decided on the division of the JNA into two armies – the Bosnian VRS and the Serbian/Yugoslav Army (VJ). This move was designed to counter accusations that Belgrade was involved in the war. However, there is compelling evidence that Belgrade continued to sponsor the Bosnian Serb forces, even during the nadir of their relationship.[12]

---

11   For more information of the Pale/Belgrade split, see: Thomas, *The Politics of Serbia in the 1990s*, 199–209. Robert Thomas, *The Politics of Serbia in the 1990s* (New York: Hurst & Co., 1999), 199–209.

12   See: Bob De Graff, 'The Wars in former Yugoslavia in the 1990s: Bringing the State Back In', in Isabelle Duyvesteyn and Jan Angstrom (eds), *Rethinking the Nature of War* (London: Frank Cass, 2005), 159–76. Further evidence from interviews confirms de Graaff's argument. According to interview evidence from leading paramilitary figures such as Arkan and Šešeslji, Belgrade continued to pay for and direct the paramilitary and regular

What then precipitated the division between the Milošević regime and their former proxies in Bosnia? As explained earlier, war's inherent violence produces a propensity to instil passion, generate hatred, and provoke hostility. This is a key ingredient of war and the effects of this hostility often form the basis for victory. The will to fight is clearly essential if one is to make policy a reality, or overcome the forces ranged against you. For Croatian and Bosnian Muslim forces, passion and the will to fight back were vital factors in thwarting Milošević's policy goal – a Greater Serbia. This can be contrasted with the incapacitating state of Serbia's desertion problem, as evident through the almost systematic refusal to adhere to official mobilization. However, in a different form the manifestation of hostility is also evident through an examination of Milošević's relationship with his proxies. As the war evolved, Milošević was unable to instil a unified policy and he found it increasingly difficult to bring events back into his control until relatively late on in the conflict.

Of course, the history of war is littered with unintended consequences, and once set in motion wars have a habit of throwing up nasty surprises and often appear to be carried along by their own momentum. In the Serbian case, the lack of cohesion reflects the extent to which there was no concrete plan. Paradoxically, just as one fatal flaw in the execution of the Serbian plan was the paucity of real passion and purpose in the majority of the Serb populace, in another way the hostility and enmity generated by early victories merged with an obdurate dislike for outside mediation, resulting in the hardening stance of the Bosnian Serb leadership when presented with pressure to toe the Milošević line.

This polarized the Serbian partnership and exacerbated an already fraught relationship. While the Bosnian Serbs wanted to join Serbia, Milošević was increasingly worried about the effect of international isolation on domestic politics. By 1993 the effects of international sanctions were beginning to bite and Belgrade sought to bring about an end to the war on favourable terms.[13] At Dayton, Milošević's first priority was neither securing agreeable borders within Bosnia-Herzegovina nor maintaining control of Eastern Slavonia. His focus was on lifting the sanctions imposed on Belgrade.[14] According to Richard Holbrooke, as the Dayton negotiations began each of the three presidents singled out their

---

forces during the division of capabilities See: LHCMA – Transcript of Interview, Vojislav Šešelj, 'Leader of Serbian Chetnik Paramilitary Organisation', *The Death of Yugoslavia* 3(69) (March 1995).

13   Following the Serb offensive at Srebrenica in 1993, the UN Security Council passed a resolution which tightened existing sanctions on Serbia; these would see Serb assets being frozen abroad and transhipments through Yugoslavia were banned. Belgrade, nor the Bosnian Serbs, planned for a protracted war. As Milošević revealed during the Dayton negotiations, 'I never thought it (the war) would go on so long.' Cited in: Holbrooke, *To End a War*, 245.

14   Holbrooke, *To End a War*, 264–5.

priorities. Holbrooke remarks, 'For Milošević, it (the priority) was sanctions'.[15] The stumbling block was that the Bosnian Serbs held considerable advantages over their rivals. Convincing them to negotiate an unfavourable peace deal was not easy. Despite sharing a mutual opponent it was not until the Washington Agreement concluded the bitter conflict between the Bosnian Muslim and Bosnian Croat forces that the strategic environment began to show any signs of changing. Set against that strategic milieu there was little to entice the Bosnian Serb leadership to come to the negotiating table. Milošević on the other hand wished to bring about a suitable end to the war, thus taking advantage of existing gains before the international and domestic political pressure had time to bear. With an eye on the bigger picture, not least his own political future, he chided the Bosnian Serbs, castigating their failure to bring about a favourable settlement to the conflict. The Bosnian Serbs, and to a certain extent the Croatian Serbs, sought to maintain their para-states with a view to incorporation into Serbia proper. In contradistinction, Milošević sought to bring about a quick end to the fighting while there was still something to gain.

With the Serbs confronting a new alliance ranged against them as a consequence of the Croat-Muslim rapprochement, new emphasis was placed on bringing about a suitable conclusion to the conflict. However, just as the Contact Group Plan in 1994 seemed to offer a resolution, Belgrade's Bosnian Serb proxies directly challenged Milošević's authority to negotiate on behalf of the Bosnian Serb parliament.[16] The schism revolving around the Contact Group Plan imitated earlier disagreements, especially those regarding the negotiations over the Vance-Owen Peace Plan. Buoyed by their early successes, the Bosnian Serbs wanted to hold out for a better deal.

Throughout 1994 there was a feeling that the international community had tired of the conflict and that it would rather be rid of the problem. In fact, the Serbs had successfully sold the conflict as essentially ethnic, where mediation was pointless. As the previous chapter highlighted, this served to stall the external powers from intervening; the UN seriously considered withdrawing its forces from the conflict during 1994. As far as Karadžić and Mladić were concerned, under these circumstances the Bosnian Serbs had little incentive to agree to the Contact Group proposal, which carved off territory already won by the Serbs in 1992. On the contrary, the Bosnian Serbs sought extended territory over and above the Contact Group offer. In the fluid and uncertain contours of the war, the Bosnian Serbs

---

15   Holbrooke, *To End a War*, 236. The Croatian priority was for the return of Eastern Slavonia. The Bosnians wanted negotiations on their Federation with the Bosnian Croats to be strengthened in order to prevent the Croat-Muslim tension unravelling at a later date – thus allowing Croatia to dominate the political fallout.

16   Ian Traynor, 'Bosnian Serbs Expected to Reject Peace Plan', *Guardian*, 19 July 1994.

very nearly realized their goal.[17] The Republika Srpska Assembly at Pale rejected the plan, precipitating the split with Belgrade and the explosion of a defamatory media campaign initiated by Milošević against Karadžić and the Pale parliament. On the eve of war in BiH in 1992, these same politicians had been cast as paladins of the Serb nation. They were now cast as 'criminals and war profiteers'.[18] Serbian propaganda, which had done so much to incite nationalism as Yugoslavia unravelled, now turned against these former Serb heroes. Milošević moved to distance himself and even withdrew material support from the Bosnian Serbs.[19] As Serbian policy diverged, seemingly spiralling out of Milošević's immediate control, the revivification of the Croats and Muslims as a cohesive fighting force again changed the contours of the conflict; putting mounting pressure on the VRS, the Croatian Army (HV) and ARBiH prevented the fulfilment of Serbian policy in Bosnia-Herzegovina. Drunk from their earlier victories, the Bosnian Serbs failed to factor in the changing international mood and the impact this would have on their sponsor and its own military position.

As the war evolved, the competing tendencies within the Trinitarian concept shaped the nature (what war is) and character (how war is fought) of the conflict. Against the odds, the ARBiH held its own against a better equipped opponent. Furthermore, either through design or desperation, the Serbian strategy of seizing territory through ethnic cleansing brought with it an unwanted media glare which crystallized opinion in the international community.[20] The end result of this was Operation Deliberate Force – NATO air strikes against Bosnian Serb positions.

Although NATO sorties signified a more emboldened international response, it was not solely the cause of Serbia's – nor Republika Srpska's – change of policy. NATO air power may have had an effect on Serbian policy, but this should not be viewed in isolation from the wider conflict. There is consensus that the precipitating cause for Serbia's willingness to come to the negotiating table was

---

17   The Contact Group Plan envisaged the separation of Bosnia into two mini-states, the Muslim-Croat Federation, which would receive 51 per cent of the territory, and Republika Srpska, which would receive the remainder. The Serbs were expected to surrender 13 towns and chunks of land won in 1992; yet at this juncture the Serbs held 70 per cent of the territory in Bosnia. The eastern Muslim enclaves of Srebrenica, Goražde and Žepa would remain in Muslim control – they would therefore also remain a strategic threat to any future Serb state.

18   Laura Silber and Alan Little, *The Death of Yugoslavia* (London: Penguin, 1995), 341.

19   Although Milošević seemingly cut-off aid to Republika Srpska, the VRS remained subordinated to the VJ and were continually re-supplied during this period. Despite being separated into three Serbian armies, the VJ, VRS, and the OS RSK (army of the Serbian Krajina), the three armies shared a common command and control system and officers frequently toured the different conflict zones. Members of the OS-RSK and the VRS continued to be paid their salaries via Belgrade. The division into separate organizations was an illusion. See: De Graff, 'Wars in former Yugoslavia in the 1990s'; and, Gow, *The Serbian Project*, 75–89.

20   Cohen, *Serpent in the Bosom*, 192–3.

more firmly linked to the battlefield successes of the Croatian HV and the ARBiH throughout 1995.[21] As President Tudjman confided to Holbrooke, Milošević understood the military situation perfectly, and adapted policy accordingly.[22] It was battle – the clearest manifestation of hostility – and especially the deteriorating position of the Bosnian Serb army that eventually brought the Serbs to the negotiating table. In the early days of the war in BiH, the Serbs' preponderance of conventional armaments had enabled the swift capture of large tracts of territory. Had the war finished then, they would have annexed around two-thirds of BiH. By 1995, however, the ARBiH was gaining in strength and had even demonstrated an ability to coordinate large-scale military operations – this was only possible as part of joint military operations with the regular Croatian army and Croat forces in Bosnia.[23]

In May 1995 in Croatia, the OS-RSK had been dealt its first major reversal of fortunes when a Croatian offensive – codenamed FLASH – swept Serb forces out of Western Slavonia. For the Serbs, FLASH was the harbinger of what was to follow when the Croats launched operation STORM. While FLASH had relatively limited aims, STORM was much more ambitious. The Croats swept into Western Bosnia, and linking with the ARBiH outflanked the Serbian forces in Knin – the capital of the Republic of Serbian Krajina.[24] Although Knin was the first target, the offensives also assisted the Bosnian breakout of the Bihać pocket, an area of Bosniak territory cut off from the SDA government in Sarajevo throughout the war. By the end of the offensives, the Serbs had conceded the whole of the Serbian Krajina, with only the sliver of territory around Vukovar still in Serb hands. In BiH, by October 1995 it even seemed that the Serbian stronghold of Banja-Luka was in danger of being lost to Croat and Bosnian Muslim forces. As Biljana Plavšić commented later, 'the moment for the liquidation of Republica Srpska was very near'.[25] Although the US urged the Croats and Bosniaks to pursue

---

21   See: Cigar, 'Serb War Effort and Termination', and, Gow, *The Serbian Project*, 241–56.

22   During negotiations at Dayton, Tudjman put pressure on Milošević by threatening military action in Eastern Slavonia and by massing Croat forces along the border. Milošević reacted by negotiating a settlement as part of the wider Dayton Accords. See: Holbrooke, *To End a War*, 260, 264–6.

23   In perhaps the best example of Croat-Muslim collusion, the ARBiH 5th Corps, in joint operations with the Croatian army, facilitated the Bosnian Muslim breakout from the Bihać pocket, forcing the VRS into a shambolic retreat which saw the loss of over 4,000 square kilometres of land in the process.

24   It appears that the OS RSK had prepared defences around Knin which would block an offensive from inside Croatia but had no preparations for a Croatian offensive operation which would sweep deep into BiH. The RSK bordered Serb-held territory in BiH and presumably thought the Croats unable to inflict any serious incursions. They were thus totally unprepared for STORM.

25   Biljana Plavšić, cited in Cigar, 'Serb War Effort and Termination', 226. Plavšić was an influential leader of the Bosnian Serb Wartime government in Pale. She was

further battlefield successes against the rattled Serb forces, the Croats brought the offensive to a conclusion. Without Croatian support, the ARBiH was unequipped to inflict serious losses on the VRS. Nevertheless, the combined arms of the Croat and Bosnian forces forcibly brought the Serbs to the negotiating table, and eventually to the Dayton Agreement.

In summation, mirroring the situation in Croatia at the end of 1991, the eventual decision to negotiate a ceasefire at Dayton in 1995 appears to have been directly linked to the ferocity of the fighting. The growing capabilities of the Croatian and Bosnian armed forces demonstrated by their battlefield successes against the VRS during 1995 forced a realignment of Serb policy. In other words, 'hostility' precipitated a Serbian policy change. HV and ARBiH gains were assisted by the international community's growing impatience with the actions of the Bosnian Serbs, especially after the atrocities at Srebrenica and Žepa, the response to which was the tougher US stance following the appointment of Holbrooke as chief negotiator. Early Serb strategy was based on the prospect of a quick victory against a far weaker enemy. As that victory failed to materialize, Serbian strategy defaulted to one which sought to hold onto existing gains in the hope that it could wear down opposition at an operational and international level.

At Dayton, a mixture of continued military pressure and high-stakes brinkmanship brought the formerly recalcitrant Bosnian Serb delegation to heel and persuaded Milošević to end the war by accepting the territorial integrity and sovereignty of Bosnia-Herzegovina as an independent state.[26] Although Republika Srpska remained extant, it did so within BiH and the Dayton Agreement prohibits political union with Serbia. Although the desire for the Serbs to secede from BiH remains, the Serbs conceded territory and prestige to gain 49 per cent of Bosnian territory; the remaining 51 per cent was conceded to the Bosnian-Croat Federation. Although there is a natural tendency to conceive of war as a linear activity with a start and an end, reality is very different and one of the most illuminating features of the Trinitarian idea is that it provides insight into the non-linear reality of conflict whereby an unordered and unpredictable series of events, and consequences of events and actions, influences the shape and character of the war. All of these threads feed into each other to form the tumultuous nature of the conflict, often dragging the war in unexpected directions. In one such example, though brutal, the tactic of ethnic cleansing Bosnian Muslims and Croats from areas of the state did make war termination a realistic possibility and until the Serbs had satisfied

---

President of Republika Srpska during 1996–98 and was charged by the ICTY in 2003 with the charges of genocide and crimes against humanity. She is now serving 11 years in a Swedish women's prison.

26   'Dayton Accords – General Framework Agreement', can be found at: http://www.ohr.int/dpa/default.asp?content_id=380 (accessed 15 May 2009). For detailed analysis of the agreement and its consequences, see: David Chandler, *Bosnia, Faking Democracy after Dayton* (London: Pluto Press, 2000); and Bose, Sumantra, *Bosnia after Dayton, Nationalist Partition and International Intervention* (London: Hurst & Co., 2002).

themselves that their territory was strategically defensible in the future they refused to negotiate. Thus, the perhaps unlikely combination of factors causing and following on from the Srebrenica massacre enabled the prospects for a negotiated settlement, both galvanizing international resolve and taking away the major stumbling block simultaneously. It is this complicated nexus between the reciprocity of cause-and-effect and policy that the Trinity helps illuminate.

Croatian Policy

The primary aim for the Croatian government at the beginning of hostilities was to secure Croatian independence within the existing republican borders. However, as the war progressed, initially in Croatia itself, but then moving into Bosnia-Herzegovina, new opportunities arose. President Tudjman was nothing if not opportunistic and while the initial policy was aimed at obviating Serbian forces from annexing Croat territory, as the war moved into Bosnia, Croatian policy altered to reflect Tudjman's desire to create a 'Greater Croatia'. This would be achieved by the annexation of Croatian populated areas of Western Herzegovina and Central Bosnia.[27] By the end of the conflict in Croatia and BiH in the autumn of 1995, Zagreb had not quite managed to accomplish all of its objectives. However, it had greatly advanced its position as a regional player. By the time of the Dayton Agreement, Croatia's territorial integrity was secure and it was a functioning state strong enough to balance Serbian interests in the region. Although Tudjman was ultimately unsuccessful vis-à-vis his policy of annexing Western Herzegovina and areas of Central Bosnia, the Croat alliance with the Bosnian Muslims and the Croat-Muslim Federation left Croatia with a great deal of political clout in Bosnia after Dayton.[28] While BiH is nominally an independent sovereign state, it is prevented from acting as such because its internal structure reflects the territorial and ethnic divisions when the war was brought to an end. To a large extent the Croatian areas of BiH owe their loyalty to Zagreb, not Sarajevo. This means that Croatia retains a huge amount of moral power in the direction of Bosnian politics.

When war began, Tudjman's policy was one of securing the territorial integrity of Croatia. Throughout the period leading to the outbreak of war and throughout the initial phases of the conflict, Croatian strategy was ill-defined and ambiguous, and its defence disordered. It is worth remembering, however, that policy originated from the tumultuous political environment which fragmented the Yugoslav state, what Clausewitz would call the 'womb of war'.[29] The nature and character of the wars of former Yugoslavia reflected the uneven and often highly confused state of political fragmentation. This is an important point and it is one reason

---

27   Gow, *The Serbian Project*, 228–9.

28   Chandler, *Bosnia, Faking Democracy after Dayton*, 66–89.

29   Clausewitz, *On War*, 273.

why Croatian policy and strategy often appeared contradictory. At the same time that the war was a source of deep insecurity, it also held out opportunities, and as Yugoslavia disintegrated neither Croatia nor its adversaries were entirely sure about what route to take.

This is of particular importance when assessing the causes of Croatia's war with Serbia; and is important also when assessing the decisions taken by the leadership as the conflict unfolded. This is not to suggest that Croatia was completely unprepared – in fact it successfully managed to prevent the JNA from realizing its aims in the republic during heavy fighting in the later part of 1991. Nevertheless, throughout this period neither Croatia's policy nor its strategy was particularly well defined and this led to confusion and conflict within Croatia's political elite, particularly between Tudjman and his chief military advisers. Although Tudjman's advisers proposed a more strident defence during the summer of 1991 Tudjman demurred. This seems partly to do with a personal reluctance to believe that the JNA would forcibly intervene in Croatia to prevent the state's independence, and partly a response to the deal brokered with Milošević regarding the partition of BiH.[30]

Clashes around the town of Knin in the Croatian Krajina had been ongoing from the middle of August 1990. Though initially small, they quickly began to escalate. At first sight they seem to highlight the popular picture of the conflict. Small bands of undisciplined Serb paramilitaries fighting it out with their Croatian counterparts in the police force offers preliminary evidence that the conflict would take on the character of a 'new war', where conventional military forces were increasingly redundant. However, Croatia had been busy re-arming its depleted military stockpiles, training new recruits, and bargaining with Serbia over the future of BiH. It was not ready for war. It was during this unreadiness, both militarily and politically, that the initial phase of hostilities took place.[31]

The largely successful disarmament of the Croatian TO in 1990 left the Croatian defence in real trouble – estimates suggest that the Croats had been left with as

---

30   Like Belgrade, Croatia's policy reacted to events and even while Croat forces repulsed the JNA and irregulars during winter 1991, Tudjman ordered the cessation of hostilities. Again, the partition of BiH seems to have been the major consideration. Martin Špegeli and Anton Tus both complained of Tudjman's inconsistency and his interference in military matters. Anton Tus, 'The War up To The Sarajevo Ceasefire', in Magaš and Žanić, *The War In Croatia and Bosnia-Herzegovina* (London: Frank Cass, 2001), 63. Tus is said to have stormed out of several meetings and successful Croatian offensives during the winter of 1991 were called off by Tudjman himself, by passing the normal chain of command and without the knowledge of his principal military advisers.

31   According to Croat Minister of Defence, Martin Špegelj, the Croats had known of the JNA plan to seize large tracts of Croatian territory. Previously intended to block the invasion of Yugoslavia by the Warsaw Pact or NATO, the so-called 'Sutjeska 2' plan, was altered to seize territory west of the Virovitica-Karlovac-Karlobag line. See: Martin Špegelj, 'The First Phase, 1990–1992: The JNA Prepares for Aggression and Croatia for Defence', in Magaš and Žanić (eds), *The War in Croatia and Bosnia-Herzegovina, 1991–1995*, 29.

little as 15,000 weapons, mostly rifles. Croatia had to embark upon an armaments programme which would enable it to resist Serbian and JNA forces.[32] This task was left primarily to Martin Špegelj, the republic's Defence Minister.[33] During the period of Croatia's re-armament programme at the beginning of the conflict, the defence was confused and under-equipped, and this may be one reason why Tudjman's policies were inconsistent. Nevertheless, by the end of May 1991, Croatia had several units operating within its borders.[34]

Rather than take the initiative militarily, as proposed by Martin Špegeli and Antun Tus, Croatian strategy was predicated upon gaining international recognition. Depicting the Croats as victims of Serbian bellicosity and ethnic intolerance underpinned this approach. Of course, international recognition of Croatian independence was by necessity a central element in the fulfilment of Zagreb's policy – the Croats simply did not have the firepower to repulse the JNA and Serb irregulars in the initial stages of the conflict. This approach formed the cornerstone of Croatia's early strategy and Tudjman spent considerable time visiting foreign capitals to drum up support, being particularly successful in Austria and Germany, Croatia's traditional sponsors.[35] Although he has since come under criticism for his complicity in the carve-up of Bosnia, this approach did provide the Croats with the time required to properly muster their military forces.

Zagreb's policy aims in the first phase of the war provide an interesting juxtaposition to its labile policy later in Bosnia. During the first stages of the war, Croatia's purpose is straightforward. It is trying to prevent territory being annexed by Serbia and the rump Yugoslavia, and is fighting to protect those Croats endangered by ethnic cleansing. As the war moved into Bosnia, there is a change of emphasis in Croatian government policy. The policy changed from one of securing territorial integrity to a more murky one which actively pursued the annexation of Western Herzegovina and parts of Central Bosnia as part of a 'Greater Croatia'. Commenting on Tudjman's wish list at Dayton, Holbrooke remarks:

> Tudjman was the critical variable. He had a clear sense of what he wanted: first, to retain eastern Slavonia; second, to create an ethnically pure Croatia; and third, to maintain maximum influence, if not control, over the Croat portion of Bosnia.[36]

---

32   'From Creation, Straight into Operation: Croatia's New Armed Forces', *Defence and Foreign Affairs*, December 31, 1992, 3.

33   Špegelj, 'The First Phase, 1990–1992', 31. Špegelj, a retired JNA General who had been in command of the 5th Army District, had been privy to JNA plans concerning the restructuring of the army throughout his tenure with the Yugoslav military.

34   Ibid., 47.

35   Markus Tanner, *Croatia: A Nation Forged in War*, second edition (New Haven: Yale University Press, 2001), 261–74.

36   Holbrooke, *To End a War*, 170.

Although, outwardly at least, the Croats and Bosnian Muslims could have acted as a bulwark against Serbian aggression and ethnic cleansing, Tudjman did not hide his appetite for expanding the Croatian state at Bosnia's expense, and he vigorously pursued this policy throughout much of the conflict in BiH. James Gow sums up Zagreb's position in BiH perfectly: it was 'either as compensation for loss of territory in Croatia itself, or, in an ideal situation, as an addition to a Croatian whole within its designated borders'. ... Gow elucidates further:

> The essential ambiguity of this position meant that, to some degree, Croatia could not lose out completely in any eventual settlement: if border changes were to be accepted, then any territory lost to Serbian forces would be compensated for by acquisition of territory from Bosnia and Herzegovina, formed around the large Croatian minority areas; and if the maintenance of existing borders were to be confirmed, then, Croatia would not gain land from Bosnia and Herzegovina but would secure the borders of Croatia itself. In an ideal world Croatia might have its cake and eat it too ... Croatia's independence within its borders was confirmed, but it also gained great influence in Bosnia and Herzegovina.[37]

Croatia's involvement in BiH can therefore be assessed in two interlinked ways. Croatian political and military intervention was part of an overarching strategy which aimed to use BiH as a bargaining chip with which to attain territorial integrity at the negotiating table. Alternatively, strategy merged with policy and thus, as Gow states above, 'Croatia might have its cake and eat it too'. Although it is far easier to assess the impact of the Trinity on such issues retrospectively, it remains a difficult task to accurately mark every twist and turn of an inherently complex series of cause-and-effect relationships. Even so, while Tudjman held the option of bargaining BiH if forced to do so, his earlier meetings with Milošević indicate his stated desire to enlarge the Croatian state – in fact, according to Stipe Mesić, it was his historical motivation.[38] As Warren Zimmermann also reveals, during one conversation with President Tudjman on 14 January 1992, Tudjman explained why Bosnia should be divided between Serbia and Croatia. Using the pretext that the Bosnian Muslims were attempting to create an Islamic state, Zimmermann recalls:

---

37 Gow, *The Serbian Project*, 230.
38 Cited in, Cohen, *Serpent in the Bosom*, 191. It was one which looked likely during the initial phases of the war in Bosnia and both Tudjman and Milošević presented their joint proposals as part of their negotiating position. 'Bosnian Partition Plan Presented at Geneva Negotiating Session', *The Washington Post*, 24 June 1993.

'they agree that the only solution is to divide up Bosnia between Serbia and Croatia'. Magnanimously, Tudjman said he didn't insist on a 50-50 division. 'Let Milošević take the larger part: he controls it anyway.'[39]

What is most clear about this situation is that Tudjman was an opportunist. Croatia's policies in relation to its own occupied territories and its involvement in Western Herzegovina and Central Bosnia indicate that Tudjman reacted to events. In other words, the momentum of the war shaped Croatian policy. Tudjman may have harboured the desire to annex territory from BiH, but he re-correlated his policy as the war dictated. Like Milošević's involvement in Croatia and in BiH, Tudjman tried to conceal his involvement in Western Herzegovina by using proxies which he could then distance himself from.[40]

Despite this, it is still possible to trace the evolution of Croatian policy through several stages. It first tried to prevent the annexation of its territory in Western and Eastern Slavonia and in the Krajina. Following partial success and the termination of the war in Croatia, Zagreb then turned its attention to BiH, and actively worked towards the annexation of Croat-dominated areas of the state, and in July 1992 Croatian forces loyal to Zagreb pronounced the creation of the new Croatian state of 'Herzeg-Bosna'.[41] As the war dragged on, and as the international community took a more robust line, Zagreb's policy altered again. Under relentless US pressure, the Croatian government signed the Washington Agreement and revivified its quasi-alliance with the Bosnian Muslims. How can this change be accounted for?

As the previous chapter revealed, the involvement of the international community throughout the conflict reflected the extent to which the external powers were beset by their own problems, and international policy regarding Bosnia remained inconsistent as a result. As also noted in the previous chapter,

---

39   Warren Zimmerman, *Origins of a Catastrophe* (New York: Random House, 1999), 182.

40   Mimicking Milošević's use of paramilitary groups, Tudjman appears to have allowed HOS (the Croat paramilitary organization – the Croatian Defence League) to operate in BiH. HOS was the biggest of the Croatian paramilitary groups and at one point had even rivalled the Croatian Army. According to Antun Tus, it was forcibly subsumed and forced into the Croatian Army by the end of 1991. Tus, 'The War up to The Sarajevo Ceasefire', 47. HOS moved into BiH as the war began, partly to escape Croatian Army control. However, by August 1992, HOS was formally incorporated into the Croatian Defense Council – which owed its allegiance to Zagreb. Interestingly, there seems to have been an anomaly in the way Croat and Serb groups were perceived and dealt with by the external powers. As argued by Sumantra Bose, although the Croats participated in wanton destruction of religious symbols, used concentration camps, and ethnically cleansed civilians, their leaders were dealt with as equals by international mediators. See: Sumantra Bose, 'The Bosnian State a Decade after Dayton' *International Peacekeeping*, 12(3) (2005), 322–35.

41   'Nationalists Proclaim New Croatian State', *Financial Times*, 6 July 1992.

this provided opportunity as well as cause for concern, and the belligerents made the most of the resulting opportunities. Therefore, the breaking of ceasefire agreements throughout the conflict should not be automatically inferred as verification of a Balkan propensity towards brutality. The inconsistency of policy emanating from the international arena produced a strategic environment in which the Balkan actors could wrestle key concessions from their opponents before international patience broke. To restate Warren Switzer's point once again, though the warring sides may have tacitly indicated approval of the international commitment, negotiation was viewed as an extension of the struggle rather than the route to peace at any cost. This resulted in speculation as to the degree of integrity and resolve of the international community, and in effect influenced how the competing belligerents viewed international involvement in the conflict. Not unsurprisingly, the ambiguity surrounding international involvement helped shape the actions of the belligerents and contributed to attempts by each side to produce their own *fait accompli*. For instance, the Croat-Muslim war that began in 1993 was partly a result of societal pressures stemming from the large influx of Muslim refugees which moved into Bosnian Croat areas after fleeing Serb atrocities. This situation was exacerbated by the Vance-Owen Peace Plan proclaimed in Geneva on 2 January 1993. The plan envisaged the division of BiH into cantons. Because the Croat-majority cantons were adjacent to Croatia itself, the plan was seen as reason to annex the entire region.[42]

At other junctures in the war, international pressure did come to bear on Zagreb, and Croatian policy was affected appropriately. Just as the Vance-Owen plan exacerbated Muslim-Croat tension in BiH, so the Washington Agreement ended that conflict. Under real pressure from the US, President Tudjman again altered course and agreed not only to bring an end to the Muslim-Croat conflict which had so paralysed the defence against the Bosnian Serbs, he also agreed to reconvene the Muslim-Croat alliance. The two warring factions now became partners in the war against the Bosnian Serbs and their masters in Belgrade. Increasingly aided by the US, the alliance offensives in Croatia and Western Bosnia were instrumental in bringing the Bosnian Serbs and Milošević to the negotiating table, and eventually to the Dayton Agreement, which brought a lasting cessation to hostilities.[43] Continued military pressure throughout the Dayton talks forced their hand.

---

42   See: Divjak, 'The First Phase, 1992–1993', 174.

43   Aware that Tudjman wanted to integrate Croatia into the Western sphere of influence, the US pressurized Zagreb by preparing the ground for a stiffening of relations with Washington. If the Croats played ball, they would gain strength by closer relations with the US. If Croatia failed to toe the line, it would suffer a similar fate as Serbia and would be ostracized from the international community – isolation which would be underpinned by harder sanctions. Galbraith proclaimed in a speech, 'Croatia has a choice of joining the West economically and politically or sharing Serbia's destiny – isolation'. See: Silber and Little, *The Death of Yugoslavia*, 322.

Interestingly, it was Croatia rather than Serbia or the Bosnian Muslims which appears to have reacted with prescience as the twists and turns of war closed down some avenues only to open other more profitable opportunities. Zagreb's policy went through several stages as the war unfolded. Focussed initially on securing the territorial integrity of the republican borders, the war opened up the chance for Tudjman to pursue a more aggressive policy aimed at annexing Croat territory in Western Herzegovina and Central Bosnia. We know also that Tudjman came under severe pressure from Redman to comply with the Washington Agreement, risking the same pariah status which was gradually paralysing Serbia. Tudjman later revealed that a variety of promises and threats were used to coerce him into acceptance of a renewed Croat-Muslim Federation.[44] As one of Holbrooke's fellow negotiators, Peter Galbraith, has since noted,

> Special Envoy Charles Redman and I put to Croatian President Tudjman: 'If Croatia would give up its ambition for a separate Croat republic in Bosnia-Herzegovina … the United States would support generous power sharing for ethnic Croats in a Muslim-Croat Federation, would support Croatia's goal of closer political relations with the West, and would work diplomatically for a political solution within Croatia's internationally recognised borders of the Serb-occupied territories.'[45]

Although the Washington Agreement and the revivified Muslim-Croat alliance closed the prospect of Croatia annexing Western Herzegovina, it also had the effect of bringing Croatia more intimately into the American political and strategic ambit. Although denied by the US at the time, Croatia enjoyed an increasingly cordial relationship with Washington. As the Dayton Negotiations took place, Richard Holbrooke recounts that on several occasions he advocated to Zagreb that Croatian forces seize as much territory as possible. In one conversation with the Croat Defence Minister, he recounts:

> 'Gojko, I want to be absolutely clear', I said. 'Nothing we said today should be construed to mean that we want you to stop the rest of the offensives, other than Banja Luka. Speed is important. We can't say so publicly, but please take Sanski Most, Prijedor, and Bosanski Novi. And do it quickly, before the Serbs regroup!'[46]

---

44   Interview with Franjo Tudjman. Croatian Television Satellite service, 24 February 1994. BBC SWB EE/1933/C, 28 February 1994.

45   Peter Galbraith, 'Negotiating Peace in Croatia: A Personal Account of the Road to Erdut', in Brad K. Blitz (ed.), *War and Change in The Balkans: Nationalism, Conflict and Cooperation* (Cambridge: Cambridge University Press, 2006), 124–31.

46   Holbrooke, *To End a War*, 167.

In a fax that Holbrooke dates as 20 September, he explains to Washington, 'If they take Sanski Most or Prijedor, both of which are in Federation hands in the Contact Group's map but which Milošević has said he will not yield in a negotiation, it would make our job easier'.[47] Although Holbrooke later expressed his bewilderment as to why Tudjman eased off the Croatian offensive before completing Washington's wish list, Tudjman's decision was probably the right one as far as Croatia was concerned. In contrast to his early political and strategic immaturity, the decision to halt the Croatian offensives at the very point when they were most powerful had far-reaching consequences. While it is clear that Croatian offensives could have gone much further, in terms of a future settlement it was likely that these gains would have benefited the ARBiH rather than the Croatians, who would have had to concede this territory at the negotiating table. Stopping the offensives at this point reminded the Bosnian Muslims of the weakness of their position and their reliance on Croatian fire-power as a force-multiplier in the war against the Bosnian Serbs. Zagreb thus had plenty of room to manoeuvre at Dayton. Croatia's alliance with the Bosnian Muslims, fraught though it may have been, brought sizeable rewards, not least in the shape of closer military and political ties with the US. Of course, Tudjman gave up his policy to annex Western Herzegovina, but his ability to identify and accept the changing political environment allowed him to change policy in accordance with the fluctuations of the unfolding war.

When the Croatian war aims are assessed following Dayton, it is evident that it succeeded in its core aim – the territorial integrity of Croatia. In fact, it was now an ethnically pure state and had thus resolved the underlying motivation for conflict in the first place. In BiH, some concessions had been made, but Croatia remains vital to the stability of Bosnia-Herzegovina and Zagreb continues to exert significant influence. In addition, one must not forget the effect of the offensives on the Serbs. Milošević was quick to concede Eastern Slavonia rather than risk a renewed confrontation with Croatia. When talks regarding Eastern Slavonia seemed to break down at Dayton, Tudjman simply stated that Croatia would retake its territory by force.[48] The Erdut Agreement, which brought about the end of Serb occupation, was signed on 12 November 1995 as part of the Dayton Agreement.[49] It provided a transition period of one year with the possibility of a further year in which the United Nations formed an administration in the region that would oversee the return of the area to Croatian control. While during the war in 1991, Croatian forces suffered from an early debilitating inferiority to their

---

47   Holbrooke, *To End a War*, 168.

48   Gregory Copley, 'Croatia Prepares For War on Eastern Slavonia', *Defence and Foreign Affairs Strategic Policy* (31 October 1995).

49   The agreement provided a transition period of one year with the possibility of a further year whereby the United Nations formed an administration in the region that would oversee the return of the area to Croatian control. Galbraith, 'Negotiating the Peace in Croatia', 129.

Serbian opponents, the situation had been transformed by the time of FLASH and STORM. The remnants of the OS RSK were no match for the Croatian HV and Milošević's desire to rid Serbia of sanctions prohibited the VJ from being dragged into any confrontation in Eastern Slavonia. By October and November, the Croatian HV was a conventional military force which, despite ongoing sanctions, was a potent regional military player.[50]

That the Croatian army had evolved from a disparate group of militias into a cohesive regional power is itself testament to the acuity of Zagreb's choices, and not a little luck, throughout the conflict. While Croatia pursued the same type of ethnically pure Croatian state that Milošević sought for Serbia, Zagreb had a better appreciation of the changing contours of the conflict and thus was able to alter Croatian goals so as to appropriate the best outcome. Like that of its opponents, Croatian policy changed, forced to do so because of the constantly changing nature of the war, itself caused by the interactivity and reciprocity of competing belligerents and the impact of hostility and uncertainty, which create the Trinity's numinous effects. However, just as policy was changed by war's nature, so too did Zagreb's reconstituted policies re-engage and alter the nature of the war.

Bosnian Policy

Of all the major belligerents engaged in the wars of former Yugoslavia, the Bosnian Muslims were the most vulnerable. At the beginning of the conflict the purpose for the Bosnian side was to preserve the territorial integrity and multi-ethnic dimension of the republic. It was faced with a better-equipped and better-prepared adversary who was in control of the tempo and character of the conflict. As it became evident that the original Bosnian policy espousing the continued multi-ethnic demographic was unachievable, a second and increasingly atavistic policy emerged – the survival of the Bosnian Muslim nation. The official policy seeking a reintegrated multi-ethnic Bosnia was eclipsed by the ferocity of the fighting. The degree to which ethnic cleansing and genocide underpinned the Serb and early Croatian policies made the prospect of a fully functioning multi-ethnic state impossible. Although the US brokered a Muslim-Croat rapprochement which provided the impetus for the Federation of Bosnia-Herzegovina (FBiH) and more pressingly for the military offensive against the Serbs during the summer and autumn of 1995, the Muslim-Croat conflict cut deeply and prohibited real movement towards a multi-ethnic state. These scars remain and are just as destabilizing as those caused by

---

50   Estimates put the Croatian force at this time at around 80,000 soldiers, with approximately 250 main battle tanks, 60,000 small arms, and 2,000 artillery pieces. Another source suggests that the HV had been supplied by Russia and had increased the number of tanks to 560, mainly through the addition of T-72s and T-55s. It also suggests that the HV had around 550 artillery pieces, 1,000 mortars and 320 armoured personnel carriers. Tim Ripley, 'Europe, Croatia's Strategic Situation', *Jane's Intelligence Review*, 1 January 1995.

Serb aggression and genocide. In a move designed to prevent the nation's survival, the Bosnian Muslim SDA progressively internalized its 'national' security and its new-found alliance with the HV remained tentative. The result of this was evolution of the ARBiH into the party army of Izetbegović's SDA.[51]

Ultimately the Bosnian Muslims failed to secure their original policy; however, they did succeed in securing the survival of the Muslim 'nation' within the Federation of Bosnia-Herzegovina. While the Bosniaks succeeded in the second of their war aims, the initial policy – securing the inviolability of Bosnia-Herzegovina's borders – was shaped and prevented as the competing, interplaying components of the Trinity collided. Just as Serbian and Croatian policy was shaped by the Trinity, so too was Bosnian policy amended and re-correlated as the conflict unfolded and the forces of hostility and chance both helped and hindered the ends-means relationship. At the beginning of the war, the Bosniaks desperately tried to involve the international community, but for much of the conflict the divisions within the international response were not consonant with Bosnian policy aims regarding a unified Bosnia-Herzegovina. Reflecting on the early British position, the government line consistently argued that the Muslim government should sue for peace because their position was hopeless.[52] The lack of international resolve provided the Serbs with an important head start until Serbian aggression peaked at Srebrenica. The assault on Srebrenica in July 1995 not only provoked international outrage, hardening the international response as a result, it also provoked the ARBiH to deepen its fragile alliance with Croatia and the Bosnian Croats when the two sides signed the Split Agreement, following which, joint HV and ARBIH operations overran the Croat Krajina and the Bihać pocket.[53] Furthermore, just as these events are products of a cause-and-effect interface, the Bosnian response was reliant on an atavistic will to survive. As Marko Attilia Hoare has argued, the failure of the combined power of the international community and the Serbs to partition Bosnia 'was due to Bosnian resistance, which thus deserves all credit'.[54] The cause-and-effect nexus creates the momentum in which the different elements of the Trinity conflate, producing its non-linear projection, and reinvesting and influencing the nature of conflict as it constantly evolves.

When war broke out in BiH at the beginning of April 1992, the Bosnian Authorities were unprepared. As Jovan Divjak, former Brigadier-General of the

---

51   Hoare, *How Bosnia Armed*, especially 102–7.

52   Brendan Simms offers a brilliantly vivid account of the British government's lack of political will in regard to Bosnian policy. Brendan Simms, *Unfinest Hour, Britain and the Destruction of Bosnia* (London: Penguin Books, 2001), 1–48.

53   'The Split Agreement' reproduced in *Bosnia Report*, No. 11, June–August 1995, 25 The agreement formally tied the HV and ARBiH into an alliance, presaging their successful joint operations in Croatian and BiH during the late summer and autumn of 1995.

54   Marko Attila Hoare, 'Discussion', *The War in Croatia and Bosnia-Herzegovina*, 303.

ARBiH, has since remarked, 'I do not know if there is another case in history of a state being internationally recognized ... without having an organized army'.[55] The lack of preparation actually reflects the extent to which Izetbegović and other Bosnian leaders had clung to the hope that there would be no repeat of the war in Croatia. In an effort not to incite trouble with the JNA and Belgrade, Izetbegović was even complicit in the removal of Bosnia TO weapons the previous year – an estimated 300,000 were returned to the JNA. Naively, the extent to which the Bosnian Muslims strived for peace left them horribly exposed to strategic setbacks throughout the war and their inability to properly prepare was a problem which they never fully overcame.[56] Although the initial disorganization of the Muslim defence gave way to a more coherent defence following the creation of the ARBiH, the Muslim forces were almost constantly on the back-foot, reacting to events rather than shaping them themselves. As Chapter 3 notes, the very fact that the Bosnian Muslim army survived at all is testament to the passion and resolve of its forces. Although Bosniak resolve eventually resulted in military stalemate with the VRS, it was reliant on external forces in order to wring policy concessions from the Serbs. Indeed perhaps most indicative of the precarious Bosniak position, the ARBiH had to align itself with the Croatian forces in Bosnia and with the Croatian regular army despite the fact they had been involved in a bitter war against each other between 1993 and 1994.

Although opinion polls prior to the elections in 1990 suggested that the majority of the population wanted to resist the nationalist forces sweeping Yugoslavia, the election divided the republic along ethnic lines. This can be explained by what Figueiredo and Weingast call 'the rationality of fear', whereby even those who wished to preserve Bosnia's multi-ethnic identity felt compelled to identify with their own groups because of the looming fear of war and accompanying insecurity.[57] When the demographics of the state are then complicated by the involvement of aggressive external influences – Croatia and Serbia, both of which attempted to annex their own chunk of Bosnian territory – the prospects of a Bosniak triumph and the realization of a multi-ethnic state were limited. In fact, that aim gradually fell away as the reciprocal effects of hostility ripped apart any prospect of a cohesive multi-ethnic state. Bosnian Muslims forces had to fight their Croat and Serb counterparts on several fronts, and also had to contend with a mini-civil war between a series of competing Bosnian Muslim factions, most notably that led by Fickret Abdić in Bihać. On the one hand, the Izetbegović regime moved increasingly to align the Muslim government to the ARBiH, and that organization evolved into a traditional party army. Furthermore, despite its

---

55   Divjak, 'The First Phase, 1992–1993', 158.

56   Markus Tanner, 'Serbian rout of Bosnian forces', *The Independent*, 28 October 1992.

57   Rui de Figueiredo Jnr and Barry Weingast, 'The Rationality of Fear: Political Opportunism and Ethnic Conflict', in Barbara F. Walters and Jack Snyder (eds), *Civil Wars, Insecurity and Intervention* (New York: Columbia University Press, 1999), 261–302.

inherent weaknesses, the ARBiH was able to crush the elements of the fifth corps of the Bosnian Army which were loyal to Fikret Abdić. While the war in BiH has been described, in the words of Mary Kaldor, as 'the archetypal example of a new war', the Bosnian Muslim government subsumed rogue elements of its own forces and crushed the Muslim rebellion. Although technologically inferior to its adversaries, it also created a conventional military force which was able to withstand Serb military offensives. Additionally, although organized crime proliferated in the Balkans, it is striking that when the Bosniak Army stabilized its position after a shaky start, it suppressed and subsumed the very groups that suggest that the war was fought by a disorganized array of militias. In one pertinent example during 'Operation Trebevic' in 1993, the ARBiH purged the 10th Mountain Division and 9th Motorised Brigades, which had formed a criminal fiefdom in Sarajevo, cementing its control of Sarajevo before exerting its strength throughout the Bosniak territories.

As the conflict in Slovenia and Croatia heralded the unravelling of the unitary state, Bosnian Muslim leaders met with Bosnian Serb delegates to reach a compromise which would preserve the unity of the republic; BiH would continue to exist in a con-federal arrangement with the rump Yugoslavia. When this arrangement broke down, the prospect of conflict increased. Unable to defend the integrity of the Bosnian state, Serbian paramilitaries – supported by the JNA – were able to seize large tracts of Bosnian Muslim territory. Major towns, particularly Zvornik and Biljennia were ethnically cleansed by Arkan's Tigers and other paramilitary groups.[58] Naively, Izetbegović still clung to the belief that war could be avoided and even asked the JNA to intervene.[59] Rather than act as the honest broker that Izetbegović hoped, the Yugoslav army merely cemented the position of the Serbian paramilitary groups by providing security from reprisal.[60] As a consequence, in the initial phase of the war in BiH, the Bosnian Muslims had no other option but to rely on a disparate range of paramilitary and militia

---

58   As early as December 1991 the JNA began taking an active role and began positioning its forces at strategically important points throughout the country. Mirroring the initial phases of the war in Croatia, the Yugoslav army had been arming Serb militias while simultaneously disarming any potential non-Serb force. Incidents where the army surrounded important centres such as Sarajevo, Mostar, and Bihać were a harbinger of what was to follow and demonstrated the seriousness of the situation.

59   Silber and Little, *Death of Yugoslavia*, 224.

60   Fickret Abdić, the then senior Muslim politician, had been prevented from reaching Bijelena by 'the Tigers'. Abdić later formed the breakaway Muslim region, the 'Autonomous Province of Western Bosnia' in the Bihać area. His mini-state and power collapsed following an ARBiH offensive against his supporters in late 1994. *Tanjung*, 8 November 1994, FBIS Daily Report (Eastern Europe), 9 November 1993, 36. Up to 1,600 Muslim fighters and as many as 5,000 civilians fled to Serb-held territory to escape the ARBiH. *Tanjug*, 12 November, FBIS Daily Report (Eastern Europe), 13 November 1991, 41.

groups.[61] Despite lacking appropriate equipment, the TO structure was reformed and alongside the Patriotic League formed the basis of the ARBiH.[62] Although Gow has highlighted the paramilitary characteristics of the early Bosnian Muslim forces, they can be differentiated from Serb groups which were armed and trained by the Federal authorities. Bosnian Muslim groups did not enjoy such material support, but they managed to resist the qualitatively superior VRS during the period 1992–95. An account of ARBiH formation and composition has previously been given; nevertheless, it is critical that the link between Bosnian resistance (hostility) and policy is made. In the starkest terms, this resistance prevented defeat and was the basis for the Bosnian Muslims' political successes, however limited they may have been.

However, just as hostility provided a bulwark against opponents and ensured survival, the vortices of conflict also prohibited the Bosnian Muslims from fulfilling their original policy aims and poisoned the idea of a multi-ethnic state. Through their inequality of fighting forces, this aim was rendered unachievable. Although the international community regularly castigated the Serbs for their role in ethnic cleansing and genocide, particularly after Srebrenica, the Bosnians had to rely on themselves for security. Plagued by internal differences, there was little cohesion in the international response. Despite international indifference, the Bosnian leadership adopted a victim strategy which consistently attempted to engineer international engagement in BiH. As noted already, the lack of a coherent international policy frequently provided opportunities for the Bosnian Muslims' adversaries.

The most salient example of course is that of the Croat-Muslim war. Although this aspect of the conflict has been highlighted already, it is worth reflecting again on how it affected Bosniak policy in relation to the unfolding nature of the conflict. Bosnian Muslim forces were tentatively allied with Bosnian-Croat forces. The breakdown of this relationship and the opportunistic behaviour of the Croats in regard to the VOPP encouraged the Croat and Serb forces to rekindle their earlier plan to partition BiH. The mixture of fear, suspicion, frustration, and hatred evident at this time was exacerbated as the flood of refugees began to arrive into Croat areas of Western Herzegovina and central Bosnia after fleeing Serb ethnic cleansing. This produced immediate societal pressures, which added to the already tense atmosphere throughout BiH. In this process we get a glimpse of the cause-effect relationship at work; of how crisis forms new problems in which belligerents of all sides must react to changing circumstances. Although Zagreb had a hand in the Croatian decision to wage war in Western and Central Bosnia, the pressure of Serb ethnic cleansing, which triggered the arrival of refugees into other areas of BIH, then caused new problems at the point of arrival.

---

61   Although the Bosnian government requested the deployment of the UN to BiH in late 1991, this was refused on the ground that the republic was not at war.

62   See: LHCMA – Transcript of Interview, Sefer Halilović, 'First Commander of Bosnian Army'. *The Death of Yugoslavia*, 3(28) (July 1994–July 1995).

Although VOPP failed, the renewed Croat-Serb relationship culminated in the Owen-Stoltenberg proposals, which would transfer territory to the warring actors on the premise that territory should be divided – Serbs 52 per cent, Bosnian Muslims 30 per cent, and Croats 18 per cent – a very close fit with the wishes of Zagreb and Belgrade.[63] Only through a mixture of ARBiH success during the war with the HVO and a renewed US effort culminating in the Washington Agreement did the Bosnian Muslims produce an effective military response. Of course, as Clausewitz states, 'the original political objects can greatly alter during the war and may finally change entirely since they are influenced by events and their probable consequences'.[64] These changes are part of war's nature and will be present to some degree in every conflict. Though it is difficult to identify the eventual effects of a given event, there are certain junctures which are especially important. For the Bosniaks the Washington Agreement is one such turning point. Events have the potential to swing things your way, the unravelling of events and their consequences can also prevent strategy from fulfilling policy. Playing the victim card had been central to Bosnian Muslim strategy and the new-found willingness to act forcibly aided the Croat-Muslim offensive during 1995. Nevertheless, the international pressure which came to bear on the participants of these Balkan wars also forced Izetbegović to accept the principles of the Dayton Agreement as the only way of preventing the Bosnian Serbs from 'compelling' the Croats and Bosniaks to their will. The Bosnian Muslims were caught trying to position themselves for a profitable outcome, but as ever they were trapped in the middle of divergent ideas, and between stronger adversaries.

Both the Vance-Owen Peace Plan and later the Owen-Stoltenberg Plan relegated the pre-war aims of the Muslim leadership. The Vance-Owen Plan effectively cantonized the state and thus the Muslim policy of maintaining a multi-ethnic state was unrecognizable; it was rejected by the SDA as a result. The Owen-Stoltenberg plan on the other hand was dictated by the Croats and Serbs and thus was highly biased against the wishes of the Bosnian Muslims. The Plan, which envisaged the partition of the state into three republics, would have ended the fighting, but it would have also ended any notion of a unified BiH, and there was no guarantee that the land-locked Muslim state would survive in the long-term. Although Izetbegovic initially reluctantly accepted the Owen-Stoltenberg plan, he would later reject it; a tactic which he regularly used and which often made him seem intransigent in the eyes of the weary international community. Ultimately of course, the Bosnian Muslims had to compromise in order to bring an end to the war and they were forced to concede policy objectives during the Washington Agreement and again at Dayton. That they were willing to concede ground here, but resist pressure at earlier junctures much less favourable to their goals, especially during the negotiations of VOPP and Owen-Stoltenberg, underlines the importance of the Bosnian Muslim motivation to fight and is

---

63   *New York Times*, 2 September, 1993.
64   Clausewitz, *On War*, 104.

testament to their indefatigability in the face of extremely bad odds. Although notoriously hard to accurately pinpoint, the interface and reciprocity of interaction produces a convergence of elements which eventually resulted in the Dayton Accords.

The Washington Agreement itself demonstrates a victory for Bosniak strategy, and is an indicator of the way in which the 'ripple' effect can have a significant impact on the course and eventual outcome of a particular war. From inauspicious beginnings, the Muslim-Croat alliance, forged out of the maelstrom of a bitter conflict, played a critical part in the eventual security of the Bosniak nation. Rivals in Western Herzegovina and Central Bosnia, these two sides put aside their differences – under US pressure – and repulsed the VRS from significant tracts of territory; driving the Serbs out of Serbian held regions of Croatia and allowing the Muslim breakout from the Bihać-pocket. US engagement was itself the result of several interlinking factors and the agreement serves as a good example of how intersecting policies converged to shape the nature of the conflict as it unfolded. It is also a good example of how events can cause effects in unintended and often unexpected ways. Although the Bosniak leaders' antipathy towards their Croat counterparts of Herceg-Bosna was well known, the SDA leadership was compelled by events to take some risks. The success of the Washington Agreement alleviated the Bosniaks' poor strategic position and provided the opportunity for the ARBiH to work in conjunction with the better-equipped and better-trained regular Croatian Army. Arms shipments making their way through Croatian territory were also no longer 'taxed' by the Croat HVO, allowing the Bosniaks to gain some degree of parity with the VRS.[65] Perhaps more importantly still, the agreement provided landlocked Bosnia with access to the sea and a link to the outside world – UN and NGO aid could now reach remote Bosnian enclaves with more ease, tactically strengthening the Bosnian position as a result.

Despite a massive disparity in terms of fighting forces, the Bosnian Muslims succeeded in securing the Muslim Nation. Although post-Dayton Bosnia is not a fully functioning state, the very fact that there is a Bosnian State at all can be credited to Bosniak resolve despite often terrible odds. Nevertheless, the Izetbegović regime had several successes. It had moved gradually away from the notion of a multi-ethnic Bosnia-Herzegovina, largely as a defensive mechanism designed to keep the Croats at arm's length, and had successfully crushed a mini-civil war in the Bihać area. Indeed, it managed to absorb paramilitary organizations which for much of the conflict were working under the control of the SDA leadership. Although earlier attempts to bring an end to the fighting had failed, the shuttle diplomacy undertaken by Holbrooke and his team, and especially the involvement of the US, eventually resolved the many conflicting issues which had sustained the

---

65   See: Cees Wibes, *Intelligence and the War in Bosnia 1992–1995* (London: Lit, 1994), chapter IV. See also: Susan Woodward, *Balkan Tragedy: Chaos and Dissolution after the Cold War* (Washington, D.C.: The Brookings Institute, 1995), 314–15.

conflict for so long. Although the original Bosnian Muslim policy was abandoned, the very fact that a Bosnian Muslim state survived at all reveals much about the twists and turns of war. What is not in doubt is the fact that when the policy of any of the belligerents enters the maelstrom of real war, war's sheer complexity obfuscates easy resolution. Like their Serb and Croat opponents, the policy of the Bosnian Muslims was shaped by the complex interplay of the Trinity, by the ripple effect of events, and by the passions that war invokes. The interactivity of these elements, although undoubtedly complicated, shaped, and frequently changed policy, but so too did policy reassert itself as Clausewitz claimed it would.

In Précis – War Termination and the Confluence of Policies: The Dayton Agreement

It is difficult to identify exactly how the interactivity of the Trinity's components impact on policy. As we know already, Clausewitz himself abjured from such prescriptions. The purpose of the present chapter has been to think about the influence which hostility and chance had on policy and purpose as the war unfolded. This has been done by drawing on the discussion in the preceding chapters and then evaluating the influence of key events as they intersect. As Clausewitz was well aware, any such examination will inevitably prove incomplete. While this is clearly the case, by examining key events against the wider context of war, one can attain a sense of the complexity of the phenomenon and appreciate the unpredictable and labile vortices of interaction which make it so fascinating and inherently dangerous. Nonetheless, in conclusion to the chapter, it is useful to close by juxtaposing the policies of the combatants as the war was brought to a close in 1995 against the original purpose when the war began in 1991.

Although the Dayton Accords were specifically about ending the conflict in BiH, the war in Bosnia was a microcosm of a wider Balkan conflagration and thus the war in BiH is not to be treated as a simple civil war between competing Bosnian nationalities. The war was started by and finished by outside actors – principally Serbia and Croatia – and was a phase of a wider conflict. Furthermore, reflecting upon the outcome of the Dayton Agreement which brought the war to a close provides an interesting comparison with the earlier policies of the warring belligerents and gives a good indication of how far policy was adapted by the opposing sides. In summary then, did policy change as the war unfolded? The simple answer to this is yes. In fact, when the original purposes of war for all the belligerent groups are contrasted against their political positions when the war was terminated, then it is clear that policy altered. At the start of the conflict in Croatia in 1991, the Milošević regime in Serbia began a campaign to create a Greater Serbia from the dissolved Yugoslavia. This policy was prevented initially by the Croats, precipitating a policy change that saw the war move into BiH. For Croatia and for the Bosnian Muslims, the initial priority was survival. In the case of Croatia, it wanted to maintain its territorial integrity as an independent state; for the Bosnian Muslim government in BiH the early purpose of the conflict was to maintain a multi-ethnic state. Although war was used as a means of bringing these varying policies to fruition, no group managed to accomplish their initial policy objectives – all parties had to alter their polices as the original motive bent to meet the realities on the ground.

As reiterated above, the initial policies of the combatants had often undergone significant alterations by the time the conflict was terminated at Dayton. The decision to implement the Dayton Agreement was a compromise position. It represents the policy option that was best for each side at that juncture, and is clearly at odds with Serbia's original motive of creating a unified 'Greater Serbia'. Neither did post-Dayton Bosnia represent the functioning multi-ethnic state that the Bosnian Muslims were seeking at the start of hostilities. Although article 1.4 of the General Framework Agreement (Annex 4) underlines the requirement for 'freedom of movement' between the separate entities, and article 2.5 stipulates the right of 'all refugees and displaced persons' to return to their homes, as noted in Chapter 4 the reality on the ground is one of separation and segregation along ethnic lines.[66] The political composition of the state reflects the end position at the close of war.

Bosnia is divided into two entities, the Federation of BIH, and Republika Srpska; there is a bicameral Parliamentary Assembly comprising the House of Peoples (with 15 delegates) and House of Representatives, which implements a Titoist style presidential rotation as a means of diffusing the latent tension in the state. The state is comprised of a complicated web of citizens' rights where 'sovereignty is devolved downwards and outwards simultaneously'.[67] Articles 3.1, 3.2 and 3.3 of the Agreement provide a framework for international and domestic responsibilities which places most power in the hands of the entities which comprise the state – these remain divided, the Federation of BiH not least.[68] While foreign policy is reserved to BiH, the armed forces and police remain divided, severely curtailing the utility of the state.[69] This means that while Serbia and Croatia can exert influence, the Bosnian Muslim situation is restricted because of military realities on the ground.

Although Zagreb conceded control to the Bosnian Federation, it had achieved most from the conflict and has managed to secure a better outcome than could

---

66   General Framework Agreement: Annex 4, 'Constitution of Bosnia and Herzegovina', articles 1.4 and 2.5.

67   Bose, *Bosnia after Dayton*, 62. Citizens of BiH are allowed to have equal membership of external states – Serbia and Croatia, which remain hugely influential.

68   Although the Federation brought the Serbs to the negotiating table, deep divisions as a result of the Muslim-Croat war in Bosnia mean that the Federation is far from united, with communities still arranged according to ethnicity.

69   For a full list of reserved and devolved areas, see: GFA (Annex 4) articles 3.1 and 3.3 'Responsibilities of Relations Between the Institutions of Bosnia and Herzegovina and the Entities'.

have been expected during the autumn of 1991, when Croatian forces seemed destined to be defeated at the hands of the JNA. The Croats secured the territorial integrity of Croatia, and had political leverage over the Bosnian Federation. As the first annex of the General Framework Agreement (1995) states:

> The Republic of Bosnia and Herzegovina, the official name of which shall henceforth be 'Bosnia and Herzegovina', shall continue its legal existence under international law as a state, with its internal structure modified as provided herein and with its present internationally recognised borders.[70]

This is not an ideal situation and this 'internal structure', in effect the Republika Srpska and the Federation, reflect the state of play at the end of 1995 and there is consensus that the international community wanted the war to end, despite the fact that the formerly warring groups were unlikely to cooperate with each other. As Izetbegović remarked at the signing ceremony, this was an 'unjust peace'.[71] For many analysts of former Yugoslavia, Dayton has done no more than hold the factions apart and there remains doubt as to whether there can be resolution and a fully functioning Bosnian state.[72] In one sense, the internal borders and the fact that Republika Srpska was extant at all may be construed as a victory for the Serbs. It seems much less of a victory in reflection of the huge losses of territory suffered at the hands of the joint HV-ARBiH offensives in the late summer of 1995. Republika Srpska does not function independently and as the General Framework Agreement prohibits secession to Serbia, in the long run its quasi-independence may prove a pyrrhic victory.[73] For all sides in the equation, the Dayton Agreement represented a compromise position and was the culmination of a complex interactive relationship between the participants of the conflict; and not least, a result of intense US political and military pressure – which was also the result of a complicated and messy process of interaction.

What is clear is that the original motives of the belligerents evolved to meet the political context of the war at the time. Policy reacted with and evolved because of the interaction of the Trinity, and in one shape or form policy had a guiding influence on the conflict at the war termination stage, just as it did during initiation.

---

70   Dayton – General Framework Agreement, 1995 (Annex 4).

71   Cited in Cohen, *Serpent in the Bosom*, 206.

72   See: Chandler, *Faking Democracy*, 135–53.

73   For further information see: GFA (Annex 4) article 3.2 'Responsibilities of the Entities'.

Conclusion

The Clausewitzian Trinity: A Framework for War in the Modern World?

The Clausewitzian paradigm has come under sustained pressure since the end of the Cold War appeared to shatter our comprehension of the traditional security and strategic environment. During this period it has become fashionable to talk about the changing nature and character of war. Deemed untraditional, these 'new wars' are thought to sit outside the purview of the Clausewitzian model. As highlighted in the introductory chapter, not only is the idea that inter-state war is obsolete premature, but the foundation of the claim that intra-state war is somehow un-Clausewitzian rests on a misinterpretation of Clausewitz's theoretical treatise, *On War*. As M.L.R. Smith so accurately points out,

> As Clausewitz above all recognised, the elemental truth is that, call it what you will – new war, ethnic war, guerrilla war, low intensity war, terrorism, or the war on terrorism – in the end, there is really only one meaningful category of war, and that is war itself.[1]

Smith's sentiments reflect a growing corpus of work which repudiates the very 'newness' of the new war thesis, and Clausewitzian scholars have done a great deal to demonstrate the longevity and universality of Clausewitzian thought. The response to the new war polemic has infused Clausewitzian scholarship with renewed vitality resulting in the rediscovery of the core Trinity – hostility, chance, and policy. In itself, this is an important point of departure from a traditional perspective which lumps Clausewitzian thought solely into a rational means-ends calculus, which consequently misinterprets the core features of Clausewitz's message in *On War*. The purpose of this book has been to use this existing scholarship as a springboard to more fully explore whether the Clausewitzian Trinity retains modern validity as an analytical device. By exploring the strengths and weaknesses of the Trinity against the trials of real war the study has also sought to determine the extent to which war is really 'an act of policy'. Although this idea has underpinned traditional explanations of strategic studies, it sits in contradistinction to Clausewitz's assertion at the conclusion of *On War*'s Chapter One, Book One that the nature of war should be understood as a 'Trinity', where policy is but one part of a tripartite model.

---

1   M.L.R. Smith, 'Guerrillas in the Mist: Reassessing Strategy and Low Intensity Warfare', *Review of International Studies*, 29(1) (2003), 34.

As highlighted in the introductory chapters, the people, army, government model which has held precedence in the traditional strategic studies literature rests on a misinterpretation of Clausewitz's message: that war is a purely rational phenomenon in which strategy is assumed to follow a neat linear trajectory from start to finish. Not only does this miss the numinous complexity of war's nature, which the Trinity is intended to explicate, it is also open to charges of obsolescence in a contemporary strategic era where inter-state war has been less prevalent. The mischaracterization of Clausewitz's Wondrous Trinity to reflect a specific rubric – the people, army, and government model – undermines the central purpose of the concept. This more widely-known model simply lacks the full explanatory power of the core concept, and consequently cannot convey the complex dynamic which Clausewitz intended to explain.

Nonetheless, while Clausewitz's modern supporters have done well to champion his core ideas, this is the first study to examine the workings, strengths, and limitations of the Trinity, and is the only study to examine the model by setting it against the experience of modern war. Clausewitz himself abjured from testing the accuracy of the concept, telling readers that to truly understand war, 'we must begin by looking at the nature of the whole; for here more than elsewhere the part and the whole must always be thought of together'.[2] Although Clausewitz comprehended war in its entirety, the parts of the whole are comprised of 'moral' and 'neutral' forces and are thought to be too elusive to explain accurately. In other words, the interaction at the heart of the Trinity is non-linear in its effects and therefore sits outside easy explication. As Clausewitz argues, 'No theorist, and no commander should bother himself with psychological and philosophical sophistries'.[3] The reason, he explains later, is that an attempt to gauge the influence of such elements can 'all too easily lead to platitudes, while the genuine spirit of inquiry soon evaporates'.[4] However, this is not particularly helpful for anyone wishing to test the modern validity of his ideas; and for the Trinity to retain its contemporary resonance it is critical to better understand the strengths and weaknesses of the concept.

In Clausewitz's conceptualization of the Trinity, he describes a model in which the 'three tendencies are like different codes of law, deep-rooted in their subject and yet variable in their relationship to one another'. He continues: 'A theory that ignores any one of them or seeks to fix an arbitrary relationship between them would conflict with reality to such an extent that for this reason alone it would be totally useless'.[5] He follows this point by arguing that to fully utilize the Trinity we must 'develop a theory that maintains a balance between these three tendencies,

---

2   Clausewitz, Carl von (1882), *On War*, translation by Michael Howard and Peter Paret (New York: Alfred A. Knopf, Everyman's Library edition, 1993), 83.

3   Clausewitz, *On War*, 158.

4   Clausewitz, *On War*, 217.

5   Clausewitz, *On War*, 101.

like an object suspended between three magnets'.[6] This is an explanatory model, which seems to reveal the true and complex nature of conflict. However, the claim that we must comprehend war as composed of three equal parts sits in contradistinction to his more famous maxim that war is the continuation of policy/politics. Determining the accuracy of the Trinity's compound is essential if we are to comprehend the true nature of war, and build on Clausewitz's claims as a platform for theorizing in our own period.

Findings

As Clausewitz was aware, the interaction at the heart of the Trinity is impossible to accurately reveal. The very point of the model is to summon a visualization of interplay and reciprocity – war stochastically stimulated by its own interaction. Of course, Clausewitz tells us that we should not construct an 'arbitrary' relationship between the component parts. Such an approach, he contends, would misinterpret the true nature of war. As a consequence, the method employed in the preceding study has been somewhat at odds with Clausewitz's own elucidation of interaction. However, as Clausewitz was cognizant, non-linearity precludes easy articulation and the Trinity is the exemplification of this complexity – explaining every 'ripple' and its multiple effects is ultimately impossible. Rather than attempt the impossible, the book has divided the Trinity into its constituent parts and explored each element singularly. While this method cannot recreate the myriad linkages of real war, it does provide a platform from which to evaluate the importance of each component. After exploring the influence of hostility and chance, the study proceeded to investigate the role of policy. Although hostility and chance are non-linear in their effects, by drawing on chapters three and four it was possible to better gauge how the first two elements of the Trinity influenced policy. Rather than charting the diplomatic and political twists and turns of the conflict, the book sought to explore the role of policy by assessing the impact of hostility and chance on key events. While certainly imperfect, by examining the concept against key examples the study has been able both to evaluate the importance of the separate elements of the concept, and then assess their importance within the concept as a whole.

By testing the concept against the real war examples of the Balkan conflicts in Croatia and Bosnia-Herzegovina, the study has underlined the importance of interactivity, which is at the heart of the Trinity and reconnects Clausewitz's idea to wars which do not fit the 'traditional' model. In addition to this, although the study has illustrated the dynamism of the Trinitarian idea, it also suggests that policy retains a guiding influence on the course and outcome of war. Though subtle, the difference is crucial and is an important point of departure from existing scholarship on the subject. The argument here is that while policy is shaped and contorted by

---

6   Clausewitz, *On War*, 101.

the pressure of real war, the feedback between war and the political communities that are waging it means that there is a guiding influence and purpose to conflict. As the Trinity exemplifies, policy is deeply affected by its own interaction with the forces in the triad. Policy is affected by unpredictability and the passion and the hatred of real war, but it remains essentially a political consideration, however messy the reality of war may be in the real world. This was the case in the wars of former Yugoslavia and it offers an updated conceptualization of the Trinity which more firmly fits Clausewitz's other arguments in *On War*.

Hostility – 'Passion, Hatred, and Enmity'

As the chapters on hostility and chance have highlighted, these elements influence war in a variety of ways. In terms of hostility, there can be little doubt that Serbian ultra-nationalism, however inspired, helped underpin the conflict with an acute bellicosity which in turn sparked resentment and a reciprocal cause-and-effect process in the other Yugoslav republics. That the conflict displayed these existing undercurrents does not mean that the combatants were irrational; it did, however, make the prospect of hostility escalating throughout the war more likely. Certainly, it was the reciprocity of war once in motion that so transfixed Clausewitz and brought him to conceptualize the Trinity. As Clausewitz predicted, in the wars in Croatia and Bosnia-Herzegovina these underlying currents of hostility and passion conflated when they interacted with opposition, and were then caught in a cyclical pattern of cause-and-effect which drove the conflict by its own momentum.

Displayed in a different way, the value of hostility is revealed also in the requirement to fight and die for one's cause – hostility is a crucial motor for combat motivation. Like hostility generally, 'passion' and the will to resist one's opponent are traits that are hard to quantify. The way in which passion is exhibited differs from person to person, group to group, and will vary according to context. It is nonetheless an essential element of conflict. Quite literally, if they are going to survive the vicissitudes of war, then combatants must be able to draw on reserves of courage. They must be prepared to fight. As the experiences of war in Croatia and Bosnia-Herzegovina convey, there is a causal link between 'purpose', means, and a willingness to invest in blood.

Despite pursuing a nationalist project to create an enlarged Serbian state, the Serbs suffered from debilitating poor morale and a systemic inability to fill the draft, which sits uneasily with the idea that these wars were the result of an ancient lust for revenge. With the exception of the paramilitary organizations – which were partly a response to disastrously poor mobilization in 1991 – the Serbs demonstrated an unwillingness to fight. This stands in stark contrast to their adversaries. Although the experiences of each group differed in important respects, both the Croats and Muslims managed to obviate the Serbs partly through a willingness to fight, despite poor odds for much of the conflict. As Clausewitz comments,

> An army that maintains its cohesion under the most murderous fire; that cannot be shaken by imaginary fears and resists well … whose physical power … has been steeled by training and privation and effort; a force that regards such efforts as a means to victory rather than a curse on its cause … such an army is imbued with true military spirit.[7]

Although the ARBiH started from the most disadvantageous position and lacked any kind of operational ability until the last months of the war, it still produced a military stalemate. The reasons for this are complex and they are interlinked with an array of factors. In what Stathis Kalyvas calls 'non-conventional symmetry', the ARBiH held its own and prevented the disappearance of the Bosnian state.[8] Unlike the Serb position, the problem was not too few fighters, but rather how to incorporate large numbers of recruits into a cohesive whole when state and military institutions lacked the infrastructure to cope. There are undoubtedly many reasons why people choose to fight. However, Serbia's territorial aggrandizement – and the tactics it employed – provoked an atavistic response from its opponents. In a very real way, while the Serbs were fighting for a Greater Serbia, the Croats in the critical early stages of the war and the Bosnian Muslims throughout the war were fighting for their very survival. Unable to win a quick victory, as the war dragged on these factors became centrally important and key ingredients in the evolution of the nature of the conflict. Clausewitz states:

> The spirit and other moral qualities of an army, a general or a government, the temper of the population of the theatre of war, the moral effects of victory or defeat – all vary greatly. They can moreover influence our objective and situation in very different ways.[9]

Although Clausewitz does much to highlight and assess the role of morale and fighting spirit, there is a tension between his view of the spirit of regular soldiers on the one hand, and irregular on the other. While regular soldiers show discipline in the face of danger, insurgencies according to Clausewitz are different, and 'in national uprisings and people's wars [its] place is taken by natural warlike qualities'.[10] Had Clausewitz lived to finish his work he may have revised this attitude. In Clausewitz's age, as in our own, such classifications can easily blur, and it is evident from the experiences in former Yugoslavia that a people's war evolved into something more closely resembling regular 'conventional war'. What is not

---

7   Clausewitz, *On War*, 220.
8   See: Kalyvas, 'Warfare in Civil Wars', 88–108.
9   Clausewitz, *On War*, 216.
10 Clausewitz, *On War*, 221.

in doubt is how the combinations of hostility highlighted by Clausewitz in the Trinitarian formula can be manifest in unexpected ways. The 'heart and temper of a nation' can have positive as well as negative effects on the course of war, acting, as Clausewitz was aware, as a force multiplier able to turn the course of events. He notes wryly, 'small things depend on great ones, unimportant on important, accidents on essentials'.[11]

Chance and Uncertainty

Although this component is the most elusive element of the triad, it is a critical aspect of war's nature and has the potential to complicate the strategies of combatants at almost every turn. As Clausewitz tells us, 'Chance makes everything more uncertain and interferes with the whole course of events.'[12] By including chance in his three-part theory, he was introducing the unpredictable; an element that by its singular nature is voluble.

It is unquestionably true that chance and uncertainty are not exclusive to war. They are a natural element in every aspect of life and weave their way through every aspect of our entire social existence. Thus, in one way the inclusion of chance simply reflects Clausewitz's understanding of the relationship between war and the rest of society. War is susceptible to chance just like any other human activity. This simple observation remains critical. During Clausewitz's era it represented a break with contemporary strategic theorists who put their faith in mathematical formulas as a way of overcoming war's unpredictability. A purely scientific, mathematical approach to war may remain a panacea for some; however, refusal to recognize war's inherent complexity too often places the focus on technology. This fails to appreciate context; reminding policymakers and strategists that war is bound by the peculiarities of uncertainty which engulf the real world is an important step in formulating a cohesive strategy. War is part of a wider political and social world, not separate from it. However, when one adds reciprocity into the equation the problems of chance and uncertainty take on a new importance. It is this interaction which is the key to understanding the Trinity and the place of chance within it.

---

11   Clausewitz, *On War*, 720.

12   Clausewitz, *On War*, 117. While in book two chance is the 'intruder', which 'interferes with the whole course of events', in the Trinity (and thus the finished book one) chance represents uncertainty. It is interlinked to his understanding of psychology and his adherence to the notion of genius, it also displays a positive side, 'chance' provides the arena where 'the creative spirit is free to roam'. This point is also made by: Herbig, 'Chance and Uncertainty', 95–116. This appears to be an acknowledgement to Machiavelli and to Thucydides, both of whom highlighted the profitable side of chance. Machiavelli proffers: 'men are able to assist Fortune but not thwart her. They can weave her designs but cannot destroy them.' See Machiavelli, *The Discourses*, vol. 1, 408.

Although chance plays an important part in all aspects of life, its pervasive power is multiplied by the process of constant interaction. War magnifies the worst aspects of chance, in the process producing a non-linear strategic environment unsuited to mathematical solutions. As Clausewitz proffers, although everything seems simple, 'the simplest thing is difficult'.[13] His inclusion of chance in his theory of war reflects much more than a simple acknowledgement of the freak event – the loss of documents that then fall into the wrong hands, or the unfortunate turn of bad weather which slows an offensive, providing one's opponents with essential time to prepare. For Clausewitz, chance equals uncertainty. As he reflects in Book Three, Chapter Seven:

> The reader expects to hear of strategic theory, of lines and angles, and instead of these denizens of the scientific world he finds himself encountering only creatures of everyday life.[14]

Chance inheres to all wars and requires dexterity, intuition, sharpness of mind, and the *coup d'oeil* of the 'genius', which Clausewitz claimed could overcome the vicissitudes of war's unpredictability. Although Clausewitz goes to some length to highlight the role of chance, incorporating this element not least in his Trinity, it is a part of war that can be overcome and he provides an entire chapter on the role of genius, and devotes the remainder of his treatise to explaining important insights into winning war. In any war, uncertainty and friction erode the strategic vision of each competitor, providing opportunities as well as difficulties should those directing war have the nerve to take them. The wars in Croatia and Bosnia-Herzegovina were no different.

Like any war, the conflict in Croatia and Bosnia-Herzegovina was engulfed by uncertainty and this was a major contributor to the character of the war. As demonstrated in the case study, chance can have multiple outcomes which preclude prediction. From the decision to use war as a tool of policy, the combatants were subject to the corrosive element of uncertainty and each had to engineer its own path through the machinations of its competitors. Of course, chance can be a help as well as a hindrance, and as Clausewitz is keen to explain, the right commander is able to overcome war's uncertainty and use it as a vehicle to strategic success. The commander able to identify the correct path out of the tumultuous complexity which surrounds him has an advantage over his competitor.

As the chapter on chance highlighted, in the final analysis it seems that the Croats and Bosnian Muslims dealt better with the unpredictability of this aspect of war. Zagreb in particular seems to have been best placed to profit from making the right choices in the unfolding and evolving conflict. Of course, it is important to point out that the importance of chance is bound to interaction and for that reason

---

13   Clausewitz, *On War*, 138.
14   Clausewitz, *On War*, 227.

should be thought of as an integral and normal component of war. It makes sense only as part of the 'whole'.

Policy

Although Clausewitz is most famous for his articulation of the idea that war is the continuation of politics, in the peroration to *On War*'s opening chapter he appears to supplant that idea with the countervailing argument that once war begins, policy is subject to the vicissitudes of reciprocity, thus losing its dominance while caught in a permanent struggle within the Trinity. The prevailing assumption is that policy/purpose is equal in importance to the other parts of the concept. As the case study highlighted, this claim is misleading and does not accurately reflect the course of the wars in Croatia and BiH. The difference is subtle, but critical. As the findings of the case studies suggest, while the original purpose (policy) is affected by war, it is not held hostage to the interplay of the Trinity's constituent parts. Clearly, as was the case for the combatants in the Balkan wars of the 1990s, the interaction may have a profound effect on policy, shaping it to encapsulate altered aspirations; but policy retains its dominant role. Because there is a continual process of feedback between the nature of war and the political context and communities which are engaged in conflict, war continues to be first and foremost a continuation of politics.

During the wars in Croatia and Bosnia-Herzegovina each of the belligerent groups was influenced by hostility and chance and these elements sustained and shaped the nature and character of the war. Both tendencies influenced policy. Nevertheless, the findings of the case study suggest that however much these elements altered the original purpose, policy reasserted itself. Despite the importance of hostility and chance as factors which shaped the war, policy altered in accordance with the evolving nature of conflict. However crude and untidy war may be in real life, because policy is symbiotically linked to the political community it remains the 'guiding' influence that Clausewitz claimed it should. There is a constant cycle of reciprocity, but this does not mean that policy stays captive within the Trinity; the effects of war's interactive nature obviously feed back to the polity; a reconfigured policy is then fed back into the conflict. It is a fact that Clausewitz underlines: war 'is governed by political aims and conditions that themselves belong to a larger whole'.[15] This is quite different to the idea that war is the result of purely rational forces. It is axiomatic that the forces of hostility – passion, hatred and enmity – and chance and uncertainty have the potential to dramatically alter policy and the purpose of going to war in the first place. Nonetheless, if one accepts that war is purposeful, then in either victory or defeat, policy reasserts itself.

---

15   Clausewitz, *On War*, 291.

Clausewitz captures the complexity of war through his Trinity, going a long way to explicating the volatility of war's nature. However, despite the Trinity's many strengths, the concept conveys a sense of endless interaction. It does not explain why wars end, and as the Dayton Agreement perfectly illustrates, although policy changed dramatically it remained a guiding influence. The Trinity may explain war's dynamism, but pinning policy as equal to hostility and chance underestimates the guiding role of politics. In fact, even if influenced by the other parts of the triad into some sort of war termination line, policy – and therefore politics – is the guiding force. The dynamism captured by the interaction of the concept should not mask the fact that the Trinity does not exist in a vacuum. It is the product of war, initiated by policy. Whether the opposing sides in any conflict like it or not, political considerations also bring such conflicts to a conclusion.

Although the case studies have demonstrated that individually the different parts of the concept are all important and all interact with each other to produce a complex matrix of interactivity, the findings of the study suggests that policy retains a preeminent position above the other two variables. At times, the war in Croatia and BiH seemed to escalate on a daily basis, and the tactics of some fighting groups suggest that they were influenced by the same forces of passion, hatred, and enmity described by Clausewitz. Yet despite instances of often brutal violence, the ferocity of these wars is not unique to the annals of strategic history. The violence that engulfed this conflict was 'normal', and certainly not un-Clausewitzian. This does not suggest that hostility and chance do not play critically important roles within the model. Rather, we can extrapolate that while hostility and chance shape the nature of the war, these traits of conflict do not completely engulf policy, turning war into something completely alien.

Whether it is to update policy options, to grasp opportunities as they present themselves, or even set in motion a policy of war termination, perhaps brought to bear directly because of chance and hostility, in an often messy way policy remains the 'guiding intelligence and war only the instrument, not vice versa'.[16] This is a critical point, and one which goes right to the heart of Clausewitzian theory and seems to reaffirm the traditional strategic studies foundation that war is politically based. It is worth clarifying this point. It is the position of this study that war is indeed a continuation of politics, as Clausewitz claimed. Yet as the Trinity suggests, policy is often greatly influenced by its conflation with war's other tendencies. War can be in the pursuit of rational goals, but it will be constantly impacted by 'irrational' factors. In Clausewitz's opinion, for war to retain its utility, policy (reason) must overcome these irrational traits by constantly re-engaging once war has begun – that is the job of strategy.

As explained above, the findings of the case studies cast doubt on the universality of the Trinity being composed of equally important features. Although the difference is a subtle one, it is critical and has implications not just for the way we should view the nature of war, but it can also better equip scholarship

---

16 Clausewitz, *On War*, 733.

of strategic studies with new insights into war in the twenty-first century. An appreciation of political and social context forms the basis from which a better understanding of war can be gleaned.

The findings of the study do not delimit the Trinitarian idea. The model captures the essence of war's complicated nature. Clausewitz's articulation of the interaction of opposites which is at the heart of the concept is simple, but brilliantly insightful nonetheless. This may seem straightforward, but at the time of its original publication Clausewitz's analysis sat in contradistinction to the message of his contemporaries and he remains a counterbalance to those who see war as a mechanical rather than a social activity. In a slight to competing theorists, he reflects that works on the subject 'did present a positive goal, but people failed to take account of the endless complexities involved … An irreconcilable conflict exists between this type of theory and actual practice'.[17] This remains a potent rejoinder to those who continue to frame war as a purely rational, linear activity.

Clausewitz uses the obvious, even the banal, to make the point that war is uniquely difficult to accurately plan for. As an explanatory model, the Trinity seems to convey this sense of unpredictability quite well. As has been postulated already, Clausewitz's conviction that the three parts of his theory are equal must be questioned. Clausewitz writes:

> A theory that ignores any one of them [hostility, chance, policy] or seeks to fix an arbitrary relationship between them would conflict with reality to such an extent that for this reason alone it would be totally useless.[18]

For scholars like Echevarria and Strachan, policy is subsumed within the theory and is held there through the constant cause-and-effect relationship that unceasingly reshapes the nature of war. Although each war will conform to a different pattern and though the features of the Trinity are hard to measure, by juxtaposing the role of policy it is possible to determine the different influences on policy. In the case studies of this book, it has been shown that policy did move; it was shaped by war, just as policy also shaped the war in a constant cycle of feedback. This feedback – the interplay between the polities waging war and the war itself – cannot be overlooked. Indeed, as Pascal Vennesson points out, 'Clausewitz's argument is not only about politics and policy, but about polity as well'. When the Trinity is artificially removed from context – the causes of a particular war, the political milieu and the character of the combatants, then it fails to connect with the forces that gave it life.

---

17   Clausewitz, *On War*, 154–5.
18   Clausewitz, *On War*, 101.

Strengths and Weaknesses of the Trinitarian Model

Strengths

As has been highlighted both in the introductory chapter and in the case studies, the individual elements of the formula conflate in a dynamic whole. As such it conveys the sense of complexity at the heart of war. In Clausewitz's view, it is only by understanding war in its entirety, how its separate strands combine to produce a unique but universal nature, that we can start to truly plan for its execution through strategy. This was the purpose of *On War* and although the non-linear aspect of the concept precludes prescription, it is this feature which illuminates war's complex interactive nature. Clausewitz was arguing that war's nature is so labile that it should not be treated as a science, rather it is more akin to the arts and we must accept war's ungovernable traits. As illustrated, Clausewitz's theory of war is a reaction to the Enlightenment thinking of his age, particularly the argument that war displays universal rules. However, as Clausewitz notes, 'no prescriptive formulation universal enough to deserve the name of law can be applied to the constant change and diversity of the phenomenon'.[19] The Trinity explains the unexplainable. What Clausewitz is divulging to his readers is simply that each war will have these universal variables. The concept is not intended to convey the idea that each war will follow the same course. Rather, it cements the notion that each war will be different, and therefore each should be treated differently.

As a way of conceptualizing the nature of war, the Trinity offers insights which are unavailable from the rational choice models and game theory which came to dominate so much of the traditional strategic studies discourse during the Cold War. Clausewitz understood that war was decidedly more erratic than his contemporaries contemplated, and he rejected their work on the grounds that their adherence to scientific formulae massively underestimated the effect that moral and neutral factors could have on the course of war's nature. In his own words, Clausewitz notes his underlying intention to 'expose such vagaries'.[20] Clausewitz's interactive formula – hostility, chance and reason/policy is far removed from the more staid rubric of the people, army, and government model peddled by some strategists after the Vietnam War. As highlighted already, it has been widely assumed that the people, army, government Trinity was the Clausewitzian formula for strategic success. Getting the three elements to work in accordance with each other was thought to be the key to military success. Clearly, however, for anyone who has read the conclusion to Chapter One, Book One, this is not what Clausewitz said at all. Modern critics of Clausewitz reject the contemporary importance of the Trinity with the charge that it conceptualizes a theory of war which purports to hold a rational means-ends calculus for planning for and waging war.

---

19 Clausewitz, *On War*, 176.
20 Clausewitz, *On War*, 252.

They accept Clausewitz's other statement that war is the continuation of politics as proof that he believed war a wholly rational enterprise. Following from this, in a world where war is fought by an increasing variety of actors, not simply states, the rational calculus of the ends-means nexus is considered obsolete on the grounds that non-state groups are not governed by the supposed political rationality of states in the Westphalian system.

The core Clausewitzian Trinity is the antithesis of this rational pseudo-scientific approach – it is the explication of war's dynamic and inherently volatile nature. The product of innumerable inter-relationships being constantly bashed around in a vortex-like maelstrom, the cause-and-effect relationship produces instability through its consequential effects. The outcome of one's actions will have unintended consequences which cannot be planned for; such problems cannot be classified easily. Instead, Clausewitz muses, they must be seen and felt. This is a clear explication of the intangible features of war, not a bad place for strategic calculation to begin. It explains also that once war is in motion, its own unique character will force an equally unique course: the product of reciprocity. It is this reciprocal contest between opponents which is the key to understanding Clausewitzian theory.[21] Any strategy which fails to adapt, to re-correlate as the nature of the conflict dictates will be apt to lose. Good strategy is one that is able to change as the nature of war itself changes. In Book Three, Chapter Sixteen Clausewitz states that:

> Like two incompatible elements, armies must continually destroy one another. Like fire and water they never find themselves in a state of equilibrium, but must keep on interacting until one of them has completely disappeared.[22]

Although the three components were divided, it is clear that the elements of the Trinity do not remain static but are part of a compound that is constantly in flux. The first two case studies in particular highlight this escalatory nature of the cause-effect association. Hostility and chance are bound to war and the interaction of these forces once war is ongoing creates the conditions for their effects to be continually fed back into the conflict, changing the nature and character of the war in the process. As the wars in Croatia and Bosnia-Herzegovina progressed, the force of war's nature produced natural opportunities as well as difficulties, and the combatants were quick to seize on them when they had the opportunity to do so.

Of course, purpose, the degree of escalation, and the provenance of chance and opportunity, will alter war from its original purpose. Rather than conveying war as a structured rational means-ends relationship, the Trinity explains a phenomenon underpinned with irrational components. This is seen by many as the culmination

---

21   Hew Strachan, 'Dialectics of War', in Hew Strachan and Andreas Herberg-Rothe (eds), *Clausewitz in the Twenty-first Century* (Oxford: Oxford University Press, 2007), 55–6.

22   Clausewitz, *On War*, 254.

of Clausewitz's search for an objective truth about war and was the result of bringing separate ideas together under one unifying theme – politics. The purpose of Clausewitz's opus was to produce a theory of war that was universal and the core Trinity can act as an explanatory model for any war. Although the interaction of the three variables is unique in every case, the core elements remain the same over time. Because the model identifies universal elements of war it stands the test of time. All wars, in all periods, will be prone to the same problems. Only the political context and resulting interaction will differ. Each war will be bound by the interaction of the Trinity, but each war will be different, subject to 'its own limiting positions, and its own peculiar preconceptions'.[23]

Weaknesses

The message of Trinity evades easy repudiation and it is patent that it retains its efficacy as an explanatory device in the modern world. Nonetheless, just as one should laud the brilliance of the model which Clausewitz has bequeathed, so too must we consider its weaknesses. Clausewitz intended the Trinity as a reference point, a reminder to soldiers that the pursuit of strategic success can be cruelly extinguished as war throws up unexpected twists and turns. The Trinity explains an almost mystical quality and the interactions of the competing components in the formula alert readers to the danger of viewing war and strategy through a rational linear lens.

However, this does not remove certain weaknesses, both in the theory itself, and in the way that it has been articulated by the traditional strategic studies community. Of course, like other aspects of Clausewitzian theory it is true that the Trinity rarely gets studied in the way that Clausewitz intended. Although the model itself is fairly easy to understand, one has first to navigate through the rest of *On War*, take into account the unfinished state of his opus, and then wade through the added layers of complexity of previous misinterpretations which cloak the true meaning of the concept. Indeed, while recent scholarship has 'reclaimed' the Trinity, there has been an unwillingness to explore its utility or question the importance of the three components in relation to each other, despite the fact that the Trinity seems to reject the central tenet of *On War*, that war is a continuation of politics. Even when one does evaluate its saliency, it quickly transpires that two parts of the Trinity defy easy examination. This weakness is clearly outweighed by the insights which can be gleaned from it. Nevertheless, the acceptance by Clausewitzian scholars that the Trinity is composed of equal parts is misleading and undermines the political means-ends nexus at the heart of Clausewitz's work.

As this study has shown, policy is influenced by its convergence with hostility and chance, but it retains a predominant position. The fault seems partly attributable to Clausewitz, partly to his modern supporters. Clausewitz's assertion that a theory which favours any one component over the other is bound to misunderstand

---

23   Clausewitz, *On War*, 715.

the nature of war has been taken as proof that each part of the Trinity is equal. Furthermore, in a bid to remove discourse on the Trinity away from its previous delineation of people, army, government, modern scholars separate the core concept from the secondary concept. In part this is to emphasize the core Trinity and for this they should be applauded. However, as this study has shown, policy cannot be disjoined from the polity which enacted it, war originates from a political community and creating a disjuncture between the Trinity and the forces that brought it into being causes an artificial barrier between the concept and the rest of Clausewitz's treatise. While the Trinity may not be linked strictly to the three parts of society that Clausewitz assumed, it is clear that war is more than the interaction of forces. These forces do not originate in a vacuum, but are part of wider political context. Fully understanding that context will illuminate the nature of war.

Although each war takes place in a cycle of perpetual give and take, not all wars necessarily escalate. While the reasons for escalation stem from context and reciprocity, Clausewitz theorized that war would not reach its ideal type because it is governed by politics. Total war as theorized by Clausewitz is an abstraction which does not exist in the real world. Indeed, if war resulted in pure violence it would turn conflict into something alien to its nature; 'here again, the abstract world is ousted by the real one and the trend to the extreme is thereby moderated'.[24] The Trinity in its current form seems to suggest an unending struggle between the components – where purpose is caught in the maelstrom of war. This is an anomaly in Clausewitzian theory and does not fit with the rest of his teachings. In fact, Clausewitz conceptualized his real/ideal war dichotomy to illustrate the point. Ideal – absolute – war in this context is treated as an abstraction. It cannot happen, and the reason he claims it cannot happen is because such violence would negate political purpose.

By invoking the visualization of the Trinity's interactivity Clausewitz reminds us that war is fraught with innumerable uncontrollable currents. This is the point Herberg-Rothe makes when he argues that it is in the conceptualization of the Trinity that Clausewitz highlights the limitations of rationality. One concurs with his argument; however, while Clausewitz was clearly outlining the limits of policy, it is hard to believe that he was intentionally delimiting the rest of his work. Indeed, one can read into the remainder of his treatise that policy remains the guiding argument. Assessing this against the findings of the case study, it is argued here that there is a case for realigning the importance of the components in the triad. While policy is often violently shaped by the vortices of conflict, it retains a guiding influence. It is worth qualifying this point. We can surmise both from the case study and from Clausewitz's arguments in *On War* and his other historical writings that war springs from political life and that he was highlighting the many divergent forces which have the power to change events and turn victory into defeat. He was stating that war is influenced by irrational traits, not that war is irrational or futile; his tome is a guide to strategic success.

---

24 Clausewitz, *On War*, 88.

In the rush to save the Trinity from wrongful mischaracterization, scholars risk making similar mistakes to those who assumed the Trinity was a conceptualization of a theory that explained a rational calculus of inter-state war. Although it is clear that the core Trinity is intended to explain war's unpredictability, Clausewitz's own linkage of the concept to a wider political base suggests that he conceived of it as being part of a bigger whole. The Trinity explains the interactivity of war once that contest begins, but unless there is a suitable grasp of the context in which conflict takes place, then the knowledge that war is reciprocal, interactive and non-linear will count for very little. In the modern era the form of the political entities and the context within which war exists can take radically different forms and there is an immediate necessity to properly engage with contemporary security realities. This is not to suggest that we need to re-conceptualize what war actually is. Although war undergoes changes in character we must not underestimate what remains continuous from one age to the next. Rather, there is a need to be more expansive by better understanding the political context and the political actors which exist interpedently within our changing international system. As Michael Handel remarks, the idea of 'national interest' may be somewhat of a neologism, 'but men fought for their interests long before the term was invented'.[25] Although different agents fight in their interests these may not always conform to traditional strategic studies practice.[26] Clearly there are grey areas where theory is subsumed by practice, but no social theory has the power to prescribe infallible truths and the ideas of Clausewitz can continue to expound valuable insights. In a globalized world, having a better appreciation of the reciprocal and consequential effects of one's actions is a pretty valuable tool; one that we need to more successfully utilize.

The complexity that the Trinity helps illuminate may never be properly understood, but it provides the platform from which one can act, and when juxtaposed against a firm knowledge of context can arm the soldier and the scholar with an important insight from which to begin. Notwithstanding his position when articulating the Trinity, Clausewitz maintains that the foremost lesson of war is to comprehend it as a political phenomenon. As he put it, 'How could it be otherwise?'[27]

---

25   Michael I. Handel, *Masters of War*, third edition (London: Frank Cass, 2001), 394. Handel uses a comparative analysis of much earlier theorists than Clausewitz to make his point. Drawing on the work of Sun Tzu and Thucydides, he highlights the lasting idea of 'vital interests', but he does so expansively, 'interests' can be construed as anything of vital interest to the state or group.

26   Mikkel Vedby Rasmussen also highlights alternative rationalities as a better way of preparing strategy in the risk society at war. See: Mikkel Vedby Rasmussen, *The Risk Society at War: Terror, Technology and Strategy in the Twenty-first Century* (Cambridge: Cambridge University Press, 2006), 202.

27   Clausewitz, *On War*, 75.

A Theory of War for the Modern World? Reflections and Future Research

Where does this leave the study of Clausewitzian theory, and the Trinity in particular? More particularly still, how do these new insights affect our understanding of war in the modern world? Can Clausewitz's model act as a workable framework from which to better analyse war in the twenty-first century? Can it contribute to our understanding of war today? There are several conclusions which can be drawn from the study.

Although the book concurs with much of the analyses of Clausewitz's modern supporters, in a clear point of departure it argues that the present conceptualization of the Trinity creates an artificial disjuncture from the political context within which war takes place – policy continues to exert a guiding influence on the course of war precisely because policy is part of a larger political story. This does not need to delimit the explanatory power of the Trinity. Policy may hold primacy, but as Clausewitz tells us, policy is not a 'tyrant', 'it must adapt itself to its chosen means, a process which can radically change it'. The difference between a Trinity of equal parts and one that is dominated, however slightly, by one element is subtle, but it has critical effects and offers a more realistic exposition of the nature of war and the complexity inherent in it. As noted above, when the text of Clausewitz's opus is juxtaposed against empirical data there is a weight of evidence that supports revision of the concept. Clausewitz tells readers that the 'process' (war) can radically change policy; this is a reminder of the complexity and capricious traits of conflict and an articulation of war's non-linearity. As he then goes on to argue, however, 'the political aim remains the first consideration'. He continues, 'Policy, then, will permeate all military operations, and in so far as their violent nature will admit, it will have a continuous influence upon them'.[28] Policy is influenced by the interaction that the Trinity was supposed to explain, but it is not totally subsumed by it. Furthermore, the dynamism in the concept is not lost because policy remains dominant. Rather than eroding the message that Clausewitz sought to tell, it underpins the message of the Trinity – its myriad interactions – by tying it back to a political base, in the process unifying the concept to his wider claim that war is a continuation of politics. War can be Trinitarian, and a continuation of politics at the same time.

Part of the reason why the form of the Trinity has been taken for granted is that Clausewitz tells us explicitly that it is the only part of his opus which he considers completed – 'The first chapter of Book One alone I regard as finished'.[29] According to Herberg-Rothe, 'When we see Clausewitz's Chapter 1 in perspective, it becomes clear that it is the only one he really did revise'.[30] The enunciation of the concept takes place in the only chapter of the entire work that Clausewitz deemed

---

28   Clausewitz, *On War*, 98–9.

29   Clausewitz, *On War*, 79.

30   Andreas Herberg-Rothe, *Clausewitz's Puzzle* (Oxford: Oxford University Press, 2007), 93.

finished. How then should one analyse his miscalculation? He may simply have got it wrong; or failed to realize the evident contradiction. Alternatively, it may be as Jon Sumida argues, Clausewitz actually paid little attention to the Trinity. It was intended as an explanatory reference for soldiers trying to adapt to the realities of combat. Equally plausibly, it is possible that it was an oversight which would have been rectified during the revision process. As we know, *On War* is an unfinished work and it is conceivable that he would have updated his model at a later date. As Gallie has postulated, that most scholars have traditionally been resistant to this idea seems to be based on 'curiously naïve assumptions'.[31] Chapter One of Book One may well have been the only finished section of *On War*, but if he had lived to revise the entire treatise it is feasible that he would have noticed the irregularity and altered the concept accordingly.

While the established assumption, as expressed not least in the Paret/Howard translation of *On War*, contends that only Chapter One of the first Book is finished, more recent scholarship has dented this traditionally accepted view. The main proponent of reinterpreting the dates of Clausewitz's prefatory notes, Azar Gat, argues that inaccuracies in the dates of the notes are misleading. According to Gat, the notes are older than originally thought. This would have left Clausewitz with much more time before his death to work on his manuscript. The supposition to be drawn is that *On War* is much more complete than we give it credit for. It is a claim with growing support and if true would remove a major pillar in the argument that Chapter One of the initial book is the only finished segment of *On War*. This begs the questions as to whether the Trinity is the final version of Clausewitz's theory. If Gat is right then there is a strong argument that the many references to war's political nature are more important than the assertion that war is a Trinity. This seems to make more sense and fits more easily into what we know both about Clausewitz and his ideas. The claim that war is the continuation of politics is the unifying theme of his entire work, undermining it seems unlikely. In fact, as Clausewitz brings his exposition of the Trinity to a conclusion, he underscores the notion that this is a work in progress, not the finished article:

> The preliminary concept of war which we have formulated casts a first ray of light on the basic structure of theory, and enables us to make an initial differentiation and identification of its major components.[32]

Although all of these 'major' components are vital, and understanding the potential volatility of their interaction is a crucial starting point for strategy to follow, policy's position as an arm of the political community – of whatever form that takes – ensures that it is embedded in context: the polity from which the

---

31   W.B. Gallie, *Philosophers of Peace and War: Kant, Clausewitz, Marx Engels and Tolstoy* (Cambridge: Cambridge University Press, 1978), 48.

32   Clausewitz, *On War*, 101.

decision to go to war was made. Policy makes war, and only policy can bring it to a conclusion. As highlighted in the opening chapters, the disjuncture from the political unit which gives the Trinity life is misleading. The consequence of this is that, in its current form, the Trinity becomes no more than yet another prescriptive model: the nature of war as a three-part formula. This is exactly what Clausewitz opposed and although war should be conceptualized as interactive, it must remain tied to context, and that is politics. As Bernard Brodie would remind us, 'war takes place within a political milieu from which it derives all its purposes'.[33] As long as we are prepared to re-conceptualize the political and strategic environment by analysing a wider range of actors outside the traditional strategic studies framework – from inter-state conflict to terrorism – as essentially different forms of the same phenomenon, then Clausewitz has a future. The Trinity does not explain everything and cannot act as a key to certain victory, but it is a stepping-stone to a better understanding of the forces inherent in war and therefore a starting guide to strategic action.

Although inter-state war remains on the agenda, the post-Cold War world has opened up a more complicated political milieu where different forms of political community and communication have added further complexity to international life; in many cases this blurs the distinction between what is international and national. The argument here is not that state war is necessarily a thing of the past, but rather that there are a range of actors with a complex range of motivations which can wage war. All too often Clausewitzian thought remains wedded to a purely traditional strategic studies preoccupation with the state. It may be true that Clausewitz conceived of _On War_ with the state in mind, but he was searching for a universal theory of conflict which would span the totality of history, and was well aware of forms of war that did not correspond neatly with inter-state war in post-Napoleonic Europe. In short, there is no reason why we should preclude the use of Clausewitz's formula today. Although the character of war in different periods evolves, there is a complex but fundamental consistency between politics – interest of some sort – and the use of war to achieve a purpose, whatever that may be. When his ideas are reduced to a supposed rationality of the state system, then it is likely that they will prove of only temporal value. However, when one thinks of politics in more expansive terms and which cover many forms of political participation, then the Trinity can continue to act as a foundation to our strategic knowledge. The Trinity, and indeed _On War_ itself, can act as a guide to the future and a basis from which to conceptualize war in its modern setting. Clausewitz could not foretell the future, but he identified key elements in the reality of war and these continue to provide valuable insights into war's nature and a steady foundation from which to better understand the complexities of conflict today.

When Clausewitz married his idea to different sections of society he sought to explain how these features influenced the core Trinity. In the rush to save Clausewitzian theory from accusations that it expounds a rational means-ends

---

33   Bernard Brodie, _War and Politics_ (New York: Macmillan, 1973), 1.

nexus suited to a past world, adherents of Clausewitz have been too quick to disassociate the two versions of the Trinity. Although the people, army, government model has been the cause of some confusion, Clausewitz did recognize that the Trinity was influenced by society and this is something which must be remembered if we are to better understand the nature of war today. Although the Trinity is now enjoying its time in the limelight, it was not until the later part of the twentieth century that it became popular, and when it did it was misinterpreted as a model explaining war as a rational activity which was devoid of unpredictability. This was undoubtedly wrong. However, in the aftermath of the Vietnam War, the Trinity became a model of strategic success and all one had to do was align the variables correctly and victory would surely follow. By debunking this misinterpretation Clausewitz's supporters have reinvested his ideas with new life and clarity. The core Trinity of hostility/passion, chance and probability, and policy/purpose must supersede the secondary model of people, army, and government. However, this should not come at the price of a disassociation of the Trinity from its political context or from the rest of Clausewitzian theory.

As explained, at present the Trinity illuminates how hostility, chance and policy interplay. This is all fine and well, but it is expedient to comprehend the causal chain that leads to war in the first place. Each war will be different and each 'Trinity' will be shaped by the political conditions prevailing at the time. The cause of war, whether total, limited, or low-intensity, will have a direct effect on the Trinity, and it remains prudent not only to understand the complexity explained by Clausewitz's model, but also the causes of war. In other words, we must better understand the fluidity of modern political systems and the influence of this on the competing parts of the Trinity. Although exegesis, the Trinity must be connected to the motivations that lead to war – indeed, the two cannot be separated. This is important, and if Clausewitzian theory is to resonate for future practitioners and theorists, trenchant views tying his ideas to one mode of war must be overcome, and inaccuracies refined. A masterwork it may be, but Clausewitz erred like the rest of us and refining his ideas acts as an elaboration rather than a repudiation of his theory of war.

As the world searched for a new source of meaning after the tumultuous conclusion of the Cold War, Clausewitzian theory was met with invective criticism. That his ideas were inveighed in this way was not particularly unusual, his thesis has been criticized since *On War* was posthumously published by his wife, Marie. This early censure was attributable to his verbose writing style, but his heterodox ideas did little to endear him to the conservative society of nineteenth-century Prussia. This was a time for maintaining political and social stasis, not for rocking the boat. The uncomfortable political realities unlocked by the French Revolution and the following Napoleonic Wars had been opportunely subdued, for the while. By delineating the people in arms – the *levée en masse* – as a major contributory factor in the 'new' way of war practised by Napoleon, Clausewitz's tome expressed uncomfortable views; contemporary readers demurred. His ideas have come under exacting scrutiny again since the end of the Cold War. Paradoxical though it may seem, while his early readers baulked at his ideas because they explained a world of war fused with uncertainty and chance, prone to escalate, and influenced by moral and neutral factors of which we have little control, modern critics accused him of preaching war as a rational activity. In a world of varied political units, the rationality of war which the Clausewitzian universe seemed to explain was obsolete.

By examining the Trinity against the wars in Croatia and Bosnia-Herzegovina and juxtaposing that evidence against re-interpretations of *On War* this study has highlighted the valuable insights that can continue to inform a foundation to our knowledge of contemporary security and strategy questions. Of course, the case study chosen for examination investigates the role of the Trinity's constituent parts by evaluating their role against one conflict; in doing so it follows Clausewitz's belief in the merits of individual cases studies. However, as Clausewitz argued, each war should be comprehended individually and other modern wars such as Iraq or Afghanistan would clearly demonstrate their individual nuances. Indeed undertaking a Trinitarian analysis of these two ongoing conflicts may provide a working model which can better equip our forces and strategies with the foresight needed to achieve the best possible outcome. Having a better awareness of the conflation of political, cultural and religious currents and the interaction of purpose and experience provides an essential starting place for winning these wars. The point of the Trinity is both to underline the universal features that link all wars across the totality of our strategic history, and to remind us that each war will be different, with its own peculiarities and limiting positions. In essence it is a model of adaptation, a reminder of war's changing context and the need to constantly reshape strategy in a fluid and changeable arena.

In the modern world where war is fought by a range of actors, the message that each war should be comprehended as unique to its own context is a powerful one and reminds policymakers and the military that war is much more than battle. Killing one's opponents in large numbers is not always a good barometer of success, and fighting without properly linking operational activity to purpose is likely to end in strategic stalemate or defeat. Writing in the early nineteenth century, Clausewitz was cognizant of a very modern problem: the difficulty of beating an enemy driven by a different rationale which does not equate victory or defeat with material objects. As he observes, 'the prospect of eventual success does not always decrease in proportion to lost battles, captured capitals, and occupied provinces, which is something that diplomats used to regard as dogma … On the contrary, one often attains one's greatest strength in the heart of one's country, when the enemy's offensive power is exhausted'.[34] It is an observation which seems particularly prescient at a time when western militaries are fighting an enemy with a different understanding of what victory actually means.

It is all too often the case that wars are bracketed into limited classifications – limited, total, low intensity – without first understanding the context within

---

34   Clausewitz, *On War*, 258.

which they take place. This can obscure the political currents that are forming the conflict and which are more important and may result in the misuse of force. This is a problem which has been associated with the RMA idea, and especially the notion that technology can act as a panacea where better weapons systems will automatically result in victory. Our strategic history is littered with strategic setbacks of qualitatively better-equipped forces by their ill-equipped opponents.

The problem here is that naked force can all too often lead to a disjuncture from purpose and context and may inadvertently cause many more problems than it solves. This is partly a leftover from the experience of war in the twentieth century, a time when we conceptualize war as an activity of states fighting for vital interests with large 'conventional' armies. Even then the idea was flawed and the experience during the period suggests a much broader mixture of conflicts. In short, our use of force to fulfil our aims must be suited to our strategic environment and this requires much more subtlety than has been exercised in our recent strategic affairs. The Clausewitzian Trinity acts as a valuable counterpoint to those who frame war as a straightforward linear activity and as such it can help influence the way we approach war and the way in which we react to the changes that interaction will certainly throw up. Rather than being outdated, the reciprocal interplay of the Trinity is a potent reminder that the nature of war is beguiling and complex, the culmination of interdependent forces continually in flux.

From a purely Western perspective this is important, and if we are to win modern conflicts it is imperative that traditional conceptions of strategic studies do not restrict action. This was the point Audrey Kurth Cronin was making when she censured America's military and political establishment for approaching the 'War on Terror' in the same way one would fight an inter-state war.[35] As most of our forces are adapted to that end this is understandable – it is a threat that has not totally diminished and there is increasing evidence that competing asymmetric forces have not totally given up territoriality or mass, but instead adopt 'hybrid forms of conflict'. Nevertheless, Cronin's point is a good one and Western governments need to wage war in its modern setting. Although Clausewitz is most commonly associated with the state and with war between large armies, as articulated above there is no doubt that he understood not just that other modes of warfare existed, but even more presciently, that force structure had to reflect political context. In fact, in his capacity as a military reformer in the Prussian state, he opposed the formation of forces which did not reflect political reality.[36] Jumping to contemporary events, the point is equally valid. When used properly, the Trinity in particular can provide a valuable springboard for action by focusing our attention not purely on technological advancement – as preached by the RMA,

35   See: Cronin, Audrey Kurth, 'Behind the Curve: Globalisation and International Terrorism', *International Security*, 27(3) (Winter 2002–3), 30–58.

36   Carl Von Clausewitz, 'On the German Federal Army (1818)', in *Carl Von Clausewitz: Historical and Political Writings*, transl. by Peter Paret and Daniel Moran (Princeton: Princeton University Press, 1992), 304–12.

but more fully on political and cultural context, and the reciprocity of opponents trying to subvert one's aims.

This is especially relevant in terms of the ongoing conflicts in Iraq and Afghanistan; the Trinitarian model can elucidate the complex interactions of belligerents which have so complicated the experience of war in these theatres of operation. In Iraq in 2003 the US and its allies clearly demonstrated the 'conventional' superiority of Western militaries and it seemed initially that they had won a convincing and relatively straightforward victory. As we now know, this was some way off the mark. As Christopher Coker remarks, 'A crushing tactical victory on the battlefield which provokes an insurgency is no success at all'.[37] The US and its allies may have held a massive qualitative advantage over their opponents but the emphasis on speed and technology undervalued the underlying political and cultural context in which the war was fought. Although the allies appear to have approached the war as if it was a continuation of the 1991 Gulf conflict, the political situation was very different, and rather than the straightforward victory that many expected, there was instead an explosion of violence that technological fixes could do little to overcome.

There is a persistent tendency to associate war purely with battle. This confuses the means of war with war itself – the focus is on the clash rather than the motivations and purpose of the belligerents that are doing the fighting. More specifically, a major failing of this approach is the tendency to underestimate how one's opponent can undermine technological advantages and react to the unfolding conflict better than you. In Iraq, the US has been caught in a complicated layered conflict with several different actors fighting for interconnected but individual motives. It was also guilty of undervaluing the extent to which the reciprocity of real war is prone to escalate. This is one element of the conflict that the US government and military failed to adequately control – the non-linear effects of violence. Their adversaries did much better in this regard. Al Qaeda, in fact, provoked more violence by directly engineering reciprocity, targeting the civilian population as a means of demonstrating the hollowness of US assurances of security, and fostering ethnic and sectarian dissonance as a result; estimates suggest up to 150,000 civilian casualties in Iraq alone, with a peak of over 3,700 civilian deaths a month in May 2006.[38] Moreover, when the US responds with high-tech conventional weapons and house searches, this exacerbates tension and results in rising enmity towards allied soldiers on the ground. This is not to suggest that having a technological advantage is a bad thing; however, when used without a full awareness of its effects it has the potential to lead to more problems. As David Kilcullen has highlighted, in both Iraq and Afghanistan the failure to properly understand the reciprocal effects of violence has underestimated how

---

37   Christopher Coker, *War in an Age of Risk* (Cambridge: Polity Press, 2009), 113.

38   Iraq Index; The Brookings Institute, www.brookings.edu/Iraqindex, July 16, 2009, 4. US military casualties up to July 2009 stands at 4, 328. British forces have lost 179.

unintentional aspects of conflict have greatly aided the West's adversaries.[39] As he points out, the majority of those fighting allied troops in Afghanistan and Iraq are not doing so on ideological grounds, but because of a reaction against Western presence in the country, revenge for deaths to family members, or because of perceived disregard for the cultural and religious sensibilities of local populations.

Using force wrongly, or without understanding the consequences of one's actions has greatly aided the propaganda efforts of the Taliban and Al Qaeda. Certainly, all wars will generate some hostility and the point is not to suggest that war has somehow become untenable. Rather the point is that force must be used appropriately, and having a better grasp of the non-linear effects of interaction is a good starting point. The Trinity may not be perfect, but an awareness of its non-linear effects, of how complexity can drive war down unexpected roads, is a foundation from which strategy can be better formulated. Understanding the interactive effects of hostility, chance, and policy may not provide the secret formula for strategic success, but these elements are active in Iraq and Afghanistan, as they were in the Balkans. An awareness of that tumultuous interaction can arm us with the basis from which we can tailor our responses.

While a qualitative military advantage is desirable, failure to grasp the central importance of politics and strategy is likely to court defeat, and equating war with warfare can often undermine the reasons for going to war in the first place. War's reciprocal nature – that it consists of two or more belligerent actors that think independently – guarantees that an opponent will rarely play to one's strengths. The unpredictability of war ensures not just that prescription is difficult, but that prediction about the outcome of a particular war founded solely on superior technology, rather than political dexterity, will sooner or later end disastrously. By concentrating on weapons systems and their employment on the battlefield, there is always the danger that the 'tacticization of strategy' will follow.[40] Concentrating on the development of new weapons system and how they perform in battle may be necessary to keep and extend a technological advantage, but this focus masks the intricacies and complexity of war governed not by new weapons systems, but by the interaction and conflation of hostility, chance, and politics. In our globalized world this complexity has become more, not less, and the Trinity offers a sound theoretical foundation from which to comprehend modern war – which is even more complicated. Although war was once seen as relatively straightforward, our present system means that strategy and policy often merge. In the 'War on Terror' the approach needs care and a careful balance between force and reconstruction, aid, democratization – some of which may be thought of as 'policies' rather than strategy. The point is that these aspects increasingly feed into each other and this undoubtedly complicates war and how we go about waging it. While acknowledging the changes in our own political system, Clausewitzian theory and

---

39 See: David Kilcullen, *The Accidental Guerrilla: Fighting Small Wars in the Midst of a Big One* (Oxford: Oxford University Press, 2009).

40 See: Handel, *Masters of War*, 353–60.

his basic question about how one can best achieve one's ends retain a vitality which is helpful to modern theorists and practitioners alike.

Writing in 2001, Mary Kaldor argued that the war in Bosnia impacted on our consciousness in a way that no other war up until that point had done. This was a pertinent observation; war has become ever more visible and this has added a layer of complexity that Clausewitz could not have foreseen. This reflects our constant interaction in a globalized and interconnected international milieu and this complex setting has shaped war and provided opportunities for asymmetric opponents to engage with more powerful militarily foes. This reflects a partial shift in emphasis away from the battlefield and can involve an equally complicated range of actors. As Chapter 4 of this study highlighted, the complexity of war was multiplied by the impact of external powers on the conflict in former Yugoslavia, and these inevitably became part of the fabric of the war. Although this study holds that war retains a timeless quality, the new war argument that the globalization of world politics has changed the way in which the military instrument is used is pertinent and of real value to strategists pursuing victory in the wars of modernity. Globalization has resulted in the consequences of reciprocity escalating and influencing war in unintended ways.

Nevertheless, although it is absolutely imperative that we comprehend war in its modern setting and with all the complexity that is consequently inherent, a Trinity which is firmly linked to political context and to purpose can help illuminate the nuances of modern conflict, inevitably displayed in a variety of forms – from terrorism to inter-state war. Common to all of these conflicts is the interactive cause-effect relationship of Clausewitz's 'Wondrous Trinity'. In this sense at least, all wars are Clausewitzian. Understanding the interplay of these forces will better equip policy makers and strategists alike. For that reason *On War* will remain the *vade mecum* of the soldier and strategist. Clausewitz's ideas are certainly not flawless; however, they offer a platform for theorizing about modern problems. While war changes over time, the traits which exhibit continuity must not be overlooked.

Bibliography

Branimir Anzulovic, *Heavenly Serbia: From Myth to Genocide* (London: Hurst & Co., 1999).

Raymond Aron, *Clausewitz, Philosopher of War*, transl. by Christine Booker and Norman Stone (London: Routledge, 1983).

Kelly Dawn Askin, *War Crimes against Women: Prosecution in International War Crimes Tribunals* (The Hague: Martinus Nighoff Publishers, 1997).

David Axe and Steve Olexa, *War Bots: How US Military Robots are Transforming War in Iraq, Afghanistan and the Future* (Nimble Books, 2008).

Ivo Banac, *The National Question in Yugoslavia: Origins, History, Politics* (Ithaca: Cornell University Press, 1988).

Ivo Banac, 'The Politics of National Homogeneity', in Brad K. Blitz (ed.), *War and Change in the Balkans: Nationalism, Conflict and Cooperation* (Cambridge: Cambridge University Press, 2006), 30–43.

Christopher Bassford and Edward J. Villacres, 'Reclaiming the Clausewitzian Trinity', *Parameters* (Autumn, 1995).

Christopher Bassford, 'Tip-Toe Through the Trinity, or The Strange Persistence of Trinitarian Warfare'. Available at www.Clausewitz.com/CWZHOME/TRINITY/TRINITY8.htm (2007) (accessed 11 July 2009).

Christopher Bassford, 'The Primacy of Policy and the "Trinity" in Clausewitz's Mature Thought', in Hew Strachan and Andreas Herberg-Rothe (eds), *Clausewitz in the Twenty-first Century* (Oxford: Oxford University Press, 2009).

Anton A. Bebler, 'The Yugoslav People's Army and the Fragmentation of a Nation', *Military Review* (August, 1993), 38–51.

Alexander Benard, 'Lessons from Iraq and Bosnia on the Theory and Practice of No-fly Zones', *The Journal of Strategic Studies*, 27(3) (September, 2004), 454–78.

Isaiah Berlin, *The Hedgehog and the Fox: An Essay on Tolstoy's View of History* (New York: Clarion, 1970).

Richard K. Betts, 'Is Strategy an Illusion?', *International Security*, 25(2) (Fall, 2000), 5–55.

Alan D. Beyerchen, 'Clausewitz, Nonlinearity and the Unpredictability of War', *International Security*, 17(3) (1992), 59–90.

Stephen Biddle, *Military Power: Explaining Victory and Defeat in Modern Battle* (Princeton, N.J.: Princeton University Press, 2004).

Jeremy Black, *War in the New Century* (London: Continuum International Publishing Group, 2001).

Brian Bond, *The Pursuit of Victory: From Napoleon to Saddam Hussein* (Oxford: Oxford University Press, 1996).

Sumantra Bose, *Bosnia after Dayton, Nationalist Partition and International Intervention* (London: Hurst & Co., 2002).

Sumantra Bose, 'The Bosnian state a decade after Dayton', *International Peacekeeping*, 12(3) (2005), 322–35.

Jane Boulden, *Peace Enforcement: The United Nations Experience in Congo, Somalia, and Bosnia* (New York: Praeger Publishers, 2001).

Antoine Bousquet, *The Scientific Way of Warfare: Order and Chaos on the Battlefields of Modernity* (London: Hurst & Co., 2009).

Magaš, Branka, *Tracking the Break-up 1980–92* (London: Verso, 1993).

Bernard Brodie, *War and Politics* (New York: Macmillan, 1973).

Bernard Brodie, 'The Continuing Relevance of *On War*', in Clausewitz, *On War* (1882), transl. by Michael Howard and Peter Paret (New York: Alfred A. Knopf, Everyman's Library edition, 1993), 50–65.

Michael E. Brown (ed.), *Ethnic Conflict and International Security* (Princeton: Princeton University Press, 1993).

David Chandler, *Bosnia: Faking Democracy after Dayton* (London: Pluto Press, 1999).

Thomas Chapman and Philip G. Roeder, 'Partition as a Solution to Wars of Nationalism: The Importance of Institutions', *American Political Science Review*, 101(4) (2007), 677–91.

Norman Cigar, *The Right to Self Defence: Thoughts on the Bosnian Arms Embargo*, Institute for European Defence and Strategic Studies, Occasional Paper 63 (London, 1995).

Norman Cigar, 'The Serb Guerrilla Option and the Yugoslav Wars: Assessing the Threat and Crafting Foreign Policy', *Journal of Slavic Military Studies*, 17(3) (2004), 485–562.

Carl von Clausewitz, *On War*, 3 vols, translated by Colonel J.J. Graham, edited by Colonel F.N. Maude (London: K. Paul, Trench, Trubner, and Company, 1908).

Carl von Clausewitz, 'Two Letters on Strategy', in Peter Paret, *Understanding War: Essays on Clausewitz and the History of Military Power* (Princeton: Princeton University Press, 1992).

Carl von Clausewitz, (1882) *On War*, translation by Michael Howard and Peter Paret (New York: Alfred A. Knopf, Everyman's Library edition, 1993).

Carl von Clausewitz, *On War*, translation by James John Graham (1873), with a new introduction by Jan Willem Honig (New York: Barnes and Noble, 2004).

Carl Von Clausewitz: Historical and Political Writings, transl. by Peter Paret and Daniel Moran (Princeton: Princeton University Press, 1992).

Eliot A. Cohen, 'Change and Transformation in Military Affairs', *The Journal of Strategic Studies*, 27(3) (2003), 395–407.

Lenard J. Cohen, *Broken Bonds, Yugoslavia's Disintegration and Balkan Politics in Transition*, 2nd edition (Boulder: Westview Press, 1995).

Lenard J. Cohen, *Serpent in the Bosom: The Rise and Fall of Slobodan Milošević* (Boulder: Westview Press, 2002).

Roger Cohen, *Hearts Grown Brutal: Sagas of Sarajevo* (New York: Random House, 1998).

Christopher Coker, 'Globalisation and Security in the Twenty-first Century: NATO and the Management of Risk', *Adelphi Paper*, Issue 345 (Oxford: Oxford University Press, 2002).

Christopher Coker, *The Future of War* (Oxford: Blackwell, 2004).

Christopher Coker, *War in an Age of Risk* (Cambridge: Polity Press, 2010).

Paul Collier, 'Doing Well out of War: An Economic Perspective', in Mats Berdal and David M. Malone, *Greed and Grievance: Economic Agendas in Civil Wars* (Boulder: Lynne Rienner Publisher, 2000).

Ivan Čolović, *The Politics of Symbol in Serbia* (London: Hurst & Co., 2002).

Gregory Copley, 'Croatia Prepares for War on Eastern Slavonia', *Defence and Foreign Affairs Strategic Policy* (31 October, 1995).

Tony Corn, 'Clausewitz in Wonderland', Policy Review – Web Special (September, 2006). Available at http://www.hoover.org/publications/policyreview/4268401. html (accessed 22 June 2009).

Ivo Daalder, *Getting to Dayton: The Making of America's Bosnia Policy* (Washington, D.C.: Brookings Institute, 2000).

Alex Danchev, 'Review Article: The Hospitality of War', *International Affairs*, 83(5) (September 2007).

J. Donia and John V.A. Fine Jr, *Bosnia-Herzegovina: A Tradition Betrayed* (New York: Columbia University Press, 1994).

Mark Duffield, *Global Governance and the New Wars* (London: Zed Books, 2001).

Isabelle Duyvesteyn, *Clausewitz and African War: Politics and Strategy in Liberia and Somalia* (Abingdon: Routledge, 2005).

Antulio J. Echevarria II, 'War and Politics: The Revolution in Military Affairs and the Continued Antulio J. Echevarria II, War,Politics, and RMA— The Legacy of Clausewitz', *Joint Forces Quarterly* (Winter, 1995), 76–82.

Antulio J. Echevarria II, *Clausewitz and Contemporary War* (Oxford: Oxford University Press, 2007).

Rachel A. Epstein, 'NATO Enlargement and the Spread of Democracy: Evidence and Expectations', *Security Studies* (2004–5).

Theo Farrell, 'The Dynamics of British Military Transformation', *International Affairs*, 84(4) (July 2008).

Rui de Figueiredo Jnr and Barry Weingast, 'The Rationality of Fear: Political Opportunism and Ethnic Conflict', in Barbara F. Walters and Jack Snyder (eds), *Civil Wars, Insecurity and Intervention* (New York: Columbia University Press, 1999), 261–302.

Bruce Fleming, 'Can Reading Clausewitz Save Us from Future Mistakes?', *Parameters* (Spring 2004).

Colin M. Fleming, 'Old or New Wars? Debating a Clausewitzian Future', *The Journal of Strategic Studies*, 32(2) (April, 2009), 213–41.

Lawrence Freedman, 'The Transformation of Strategic Affairs', *Adelphi Paper*, 45(379) (March 2006).

V. Gagnon, 'Ethnic Nationalism and International Conflict: The Case of Serbia', *International Security* (Winter, 1994–95), 130–66.

Peter Galbraith, 'Negotiating Peace in Croatia: A Personal Account of the Road to Erdut', in Brad K. Blitz (ed.), *War and Change in The Balkans: Nationalism, Conflict and Cooperation* (Cambridge: Cambridge University Press, 2006), 124–31.

W.B. Gallie, *Philosophers of Peace and War: Kant, Clausewitz, Marx Engels and Tolstoy* (Cambridge: Cambridge University Press, 1978).

Azar Gat, *A History of Military Thought: From the Enlightenment to the Cold War* (Oxford: Oxford University Press, 2001).

James Gleick, *Chaos: Making a New Science* (New York: Viking, 1987).

Eric Gordy, *The Culture of Power in Serbia: Nationalism and the Destructions of Alternatives* (Pennsylvania: Pennsylvania State University Press, 1999).

James Gow and Cathy Carmichael, *Slovenia and the Slovenes* (London: Hurst & Co., 2000).

James Gow, *The Serbian Project and its Adversaries* (London: Hurst & Co., 2003).

Bob De Graff, 'The Wars in Former Yugoslavia in the 1990s: Bringing the State Back In', in Isabelle Duyvesteyn and Jan Angstrom (eds), *Rethinking the Nature of War* (London: Frank Cass, 2005), 159–76.

Colin Gray, *Another Bloody Century* (London: Weidenfeld & Nicolson, 2005).

Roy Gutman, *Witness to Genocide: The First Inside Account of the Horrors of 'Ethnic Cleansing' in Bosnia* (Longmead: Element, 1993).

Ian Hacking, *The Taming of Chance* (Cambridge: Cambridge University Press, 1990).

Werner Hahlweg, 'Clausewitz and Guerrilla Warfare' in Michael Handel (ed.), *Clausewitz and Modern Strategy* (London: Frank Cass, 1986), 127–33.

Thomas X. Hammes, *The Sling and the Stone: On War in the 21st Century* (St Paul: Zenith Press, 2004).

Michael I. Handel (ed.), *Intelligence and Military Operations* (London: Frank Cass, 1990).

Michael I. Handel, *Masters of War*, third edition (London: Frank Cass, 2001).

Pavlos Hatzopoulos, *The Balkans beyond Nationalism and Identity, International Relations and Ideology* (London: I.B. Tauris, 2008).

Ryan Hendrickson, 'Leadership at NATO: Secretary General Manfred Woerner and the Crisis in Bosnia', *The Journal of Strategic Studies*, 27(3) (September 2004).

Andreas Herberg-Rothe, *Clausewitz's Puzzle* (Oxford: Oxford University Press, 2007).

Katherine L. Herbig, 'Chance and Uncertainty in *On War*', in Michael Handel (ed.), *Clausewitz and Modern Strategy* (London: Frank Cass, 1986), 127–33.

Beatrice Heuser, *Reading Clausewitz* (London: Pimlico, 2002).

Marko Attila Hoare, *How Bosnia Armed* (London: Saqi Books, 2004).

Frank G. Hoffman, '"Hybrid Threats": Reconceptualising the Evolving Character of Modern War', *Strategic Forum*, Institute for National Strategic Studies, National Defense University. No. 240 (April, 2009), 1–8.

Richard Holbrooke, *To End a War* (New York: Modern Library, 1999).

Terrence Holmes, 'Planning versus Chaos in Clausewitz's *On War*', *The Journal of Strategic Studies*, 30(1) (February, 2007), 129–51.

Kalevi J. Holsti, 'The Coming Chaos? Armed Conflict in the Worlds Periphery', in T.V. Paul and John A. Hall (eds), *International Order and the Future of World Politics* (Cambridge: Cambridge University Press, 1999).

Jan Willem Honig and Norbert Both, *Srebrenica, Record of a War Crime* (London: Penguin Books, 1996).

Jan Willem Honig, 'Clausewitz's On War: Problems with Text and Translation', in Hew Strachan and Andreas Herberg-Rothe (eds), *Clausewitz in the Twenty-first Century* (Oxford: Oxford University Press, 2007).

Michael Howard, 'When are Wars Decisive?', *Survival*, 41(1) (Spring 1999), 126–35.

Dominic D.P. Johnson, 'Darwinian Selection in Asymmetric Warfare: The Natural Advantage of Insurgents and Terrorists', *Journal of the Washington Academy of Sciences*, 95(3) (2009), 89–112

Dominic D.P. Johnson and Dominic Tierney, *Failing to Win: Perceptions of Victory and Defeat in International Politics* (Cambridge, MA: Harvard University Press, 2006).

Dejan Jović, 'The Slovenian-Croatian Confederal Proposal: A Tactical Move or an Ultimate Solution?', in Lenard J. Cohen and Jasna Dragović-Soso (eds), *State Collapse in South-Eastern Europe: New Perspectives on Yugoslavia's Disintegration* (West Lafayette, Indiana: Purdue University Press, 2008), 249–80.

Tim Judah, *The Serbs, History, Myth and the Destruction of Yugoslavia*, second edition (New Haven: Yale University Press, 2000).

Tim Judah, *Kosovo, War and Revenge* (New Haven: Yale University Press, 2002).

David Kahn, 'Clausewitz and Intelligence', in Michael Handel (ed.), *Clausewitz and Modern Strategy* (London: Frank Cass, 1986), 116–26.

Chaim Kaufmann, 'Possible and Impossible Solutions to Ethnic Civil Wars', *International Security*, 20(4) (Spring, 1996), 136–75.

Mary Kaldor, *New and Old Wars* (Cambridge: Polity Press, 2001).

Stathis N. Kalyvas and Nicholas Sambanis, 'Bosnia's Civil War: Origins and Violence Dynamics', in Paul Collier and Nicholas Sambanis (eds), *Understanding Civil War: Evidence and Analysis* (Washington, D.C.: The World Bank, 2005), vol. 2, 191–229.

Stathis N. Kalyvas, 'Warfare in Civil Wars', in Isabelle Duyvesteyn and Jan Angstrom (eds), *Rethinking the Nature of War* (London: Frank Cass, 2005), 88–108.

John Keegan, *A History of Warfare* (London: Pimlico, 1994).

David Keen, 'The Economic Functions of Violence in Civil Wars', *Adelphi Paper* (London: International Institute of Strategic Studies, 1998), 320.

Giles Kepel, *Jihad: The Trail of Political Islam* (London: I.B. Tauris, 2002).

David Kilcullen, *The Accidental Guerrilla* (London: Hurst & Co., 2009).

A.J. King, D.D.P. Johnson and M. Van Vugt, 'The Origins and Evolution of Leadership', *Current Biology* 19(19) (2009), 1591–682.

Stuart Kinross, *Clausewitz and America: Strategic Thought and Practice from Vietnam to Iraq* (London: Routledge, 2005).

MacGregor Knox and Williamson Murray, 'Thinking about Revolutions in Warfare', in MacGregor Knox and Williamson Murray (eds), *The Dynamics of Military Revolution, 1300–2050* (Cambridge: Cambridge University Press, 2001).

Avi Kober, 'The Israel Defence Forces in the Second Lebanon War: Why the Poor Performance?', *Journal of Strategic Studies*, 31(1) (2008).

David A. Lake and Donald Rothchild, 'Containing Fear: The Origins and Management of Ethnic Conflict', *International Security*, 21(2) (Fall 1996), 41–75.

John Lampe, *Yugoslavia as History: Twice There Was a Country*, revised edition (Cambridge: Cambridge University Press, 2000).

Quiao Lang and Wang Xiangsu, *Unrestricted Warfare* (Bejing: PLA Literature and Arts Publishing House, 1999).

Z. Lazarević, 'Economic History of Twentieth Century Slovenia', in J. Benderly and E. Kraft (eds), *Independent Slovenia* (New York: St Martin's Press, 1994).

Rodger Lewin, *Complexity: Life at the Edge of Chaos* (London: Phoenix; new edition, 2001).

William S. Lind, 'The Changing Face of War: Into the Fourth Generation', *Military Review*, 85(5) (1989).

David Lonsdale, 'Clausewitz and the Information Age', in Hew Strachan and Andreas Herberg-Rothe (eds), *Clausewitz in the Twenty-first Century* (Oxford: Oxford University Press, 2007), 231–50.

Edward Luttwak, 'Toward Post-Heroic Warfare', *Foreign Affairs*, 74(3) (1995), 109–22.

David Bruce Macdonald, *Balkan Holocausts?: Serbian and Croatian Victim Centred Propaganda and the War in Yugoslavia* (Manchester: Manchester University Press, 2003).

Noel Malcolm, *Kosovo, a Short History* (London: Macmillan, 1998).

Noel Malcolm, *A Short History of Bosnia*, new updated edition (New York: New York University Press, 1996).

Robert Mandel, 'Defining Postwar Victory', in Jan Angstrom and Isabelle Duyvesteyn (eds), *Understanding Victory and Defeat in Contemporary War* (Abingdon: Routledge, 2007).

Michael Mandelbaum, 'Is Major War Obsolete?' *Survival* (Winter, 1998–99).

Monty Marshall, 'Systems at Risk: Violence, Diffusion, and Disintegration in the Middle East', in David Carmet and Patrick James (eds), *Wars in the Midst of Peace: The International Politics of Ethnic Conflict* (Pittsburgh: University of Pittsburg Press, 1997), 82–115.

William Martel, *Victory in War: Foundations of Modern Military Policy* (Cambridge: Cambridge University Press, 2007).

Stipe Mesić, *Demise of Yugoslavia: A Political Memoir* (Budapest: Central European University Press, 2004).

Steven Metz, 'A Wake for Clausewitz: Toward a Philosophy of 21st Century Warfare', *Parameters* (Winter, 1994–95), 126–32.

John Mueller, *Retreat from Doomsday: The Obsolescence of Major War* (New York: Basic Books, 1989).

John Mueller, 'The Banality of Ethnic War', *International Security*, 25(1) (Summer, 2000), 42–70.

Herfried Münkler, *The New Wars* (Cambridge: Polity Press, 2005).

Williamson Murray, 'Clausewitz Out, Computers In: Military Culture and Technological Hubris', *The National Interest* (Summer, 1997).

David Owen, *Balkan Odyssey* (New York: Harcourt Brace & Company, 1995).

Peter Paret, *Clausewitz and the State. The Man, His Theories, and His Times*, 2007 edition (Princeton: Princeton University Press, 2007).

Vjekoslav Perica, *Balkan Idols, Religion and Nationalism in Yugoslav States* (Oxford: Oxford University Press, 2002).

Kemal Pervanic, *The Killing Days* (London: Blake, 1999).

Ralph Peters, 'Constant Conflict', *Parameters* (Summer, 1997).

Barry Posen, 'The Security Dilemma and Ethnic Conflict', *Survival*, 35(1) (Spring 1993).

Colin Powell, with Joseph Persico, *My American Journey* (New York: Random House, 1995).

Sabrina P. Ramet, *Balkan Babel, the Disintegration of Yugoslavia from the Death of Tito to the Fall of Milošević.* fourth edition (Boulder: Westview Press, 2002).

Mikkel Vedby Rasmussen, *The Risk Society at War: Terror, Technology and Strategy in the Twenty-first Century* (Cambridge: Cambridge University Press, 2006).

Michael Roberts, 'The Military Revolution', *Essays in Swedish History* (London: Weidenfeld & Nicolson, 1967).

Donald H. Rumsfeld, 'Transforming the Military', *Foreign Affairs*, 81(3) (May/ June, 2002), 20–32.

Rafe Sagarin, 'Adapt or Die: What Charles Darwin can Teach Tom Ridge about Homeland Security', *Foreign Policy* (September/October, 2003), 68–9.

Rafe Sagarin and Terry Taylor (eds), *Natural Security: A Darwinian Approach to a Dangerous World* (Berkeley: University of California Press, 2008).

Nicholas Sambanis, 'Partition as a Solution to Ethnic War: An Empirical Critique of the Theoretical Literature', *World Politics*, 52(4) (July, 2000), 437–83.

Meredith Reid Sarkees, 'The Correlates of War Data on War: An Update to 1997', *Conflict Management & Peace Sciences* (2000), 1811, 123–44.

William A. Schabas, *The UN International Criminal Tribunals: The Former Yugoslavia, Rwanda and Sierra Leone* (Cambridge: Cambridge University Press, 2006).

Charles R. Shrader, *The Muslim-Croat Civil War in Central Bosnia – A Military History, 1992–1994* (Texas: A.&M. University Press, 2003).

Laura Silber and Alan Little, *The Death of Yugoslavia* (London: Penguin, 1995).

Brendan Simms, *Unfinest Hour, Britain and the Destruction of Bosnia* (London: Penguin, 2001).

P.W. Singer, *Wired for War: The Robotics Revolution and Conflict in the 21st Century* (New York: Penguin, 2009).

Hugh Smith, *On Clausewitz: A Study of Military and Political Ideas* (Basingstoke: Palgrave MacMillan, 2004).

Jack Snyder and Robert Jervis, 'Civil War and the Security Dilemma', in B.F. Walter and J. Snyder (eds), *Civil Wars, Insecurity, and Intervention* (New York: Columbia University Press, 1999), 15–37.

Hew Strachan, *Clausewitz's On War* (New York: Atlantic Monthly Press, 2006).

Chuck Sudetic, *Blood and Vengeance: One Family's Story of the War in Bosnia* (New York: Norton, 1998).

Jon Tetsuro Sumida, *Decoding Clausewitz: A New Approach to On War* (Kansas: University Press of Kansas, 2008).

Col. Harry, G. Summers Jr, *On Strategy: A Critical Analysis of the Vietnam War* (Novato, CA: Presidio Press, 1982).

Waren Switzer, 'International Military Responses to the Balkan Wars: Crisis and Analysis', in Magaš and Žanić (eds), *The War in Croatia and Bosnia-Herzegovina, 1991–1995* (London: Frank Cass, 2001).

Markus Tanner, *Croatia: A Nation Forged in War*, second edition (New Haven: Yale University Press, 2001).

Bradely A. Thayer, 'Bringing in Darwin: Evolutionary Theory, Realism and International Politics', *International Security*, 25(2) (Fall 2000), 124–51.

The Military Balance 1991–1992, International Institute for Strategic Studies (London: The Institute, 1992).

The Military Balance 1994–1995, International Institute for Strategic Studies (London: The Institute, 1994).

The Responsibility to Protect: Report of the International Commission on Intervention and State Sovereignty (Ottawa: International Development Research Centre, 2001).

Robert Thomas, *The Politics of Serbia in the 1990s* (New York: Hurst & Co., 1999).

Snezana Trifunovška (ed.), *Yugoslavia through Documents from its Creation to its Dissolution* (Dordecht: Martinus Nijhoff Publishers, 1994).

Mao Tse-tung, *On Guerrilla Warfare*, translated by Brig.-Gen. Samuel B Griffith (1961) (Urbana: University of Illinois Press, 2000).

Martin Van Creveld, *The Transformation of War* (New York: The Free Press, 1991).

Nebojša Vladisavljević, *Serbia's Anti-bureaucratic Revolution: Milošević, the Fall of Communism and National Mobilization* (Basingstoke: Palgrave Macmillan, 2008).

M. Waldrop, *Complexity: The Emerging Science at the Edge of Order and Chaos* (New York: Simon and Schuster, 1992).

Stephen M. Walt, 'The Search for a Science of Strategy: A Review Essay', *International Security*, 12 (Summer, 1987).

Casper W. Weinberger, 'U.S. Defense Strategy', *Foreign Affairs* (Spring, 1986).

Casper W. Weinberger, *Fighting for Peace: Seven Critical Years in the Pentagon* (New York: Warner Books, 1990).

David Welsh, 'Domestic Politics and Ethnic Conflict', *Survival*, 35(1) (Spring 1993), 63–90.

Cees Wibes, *Intelligence and the War in Bosnia 1992–1995* (London: Lit, 1994).

Michael, J. Williams, *NATO, Security and Risk Management. From Kosovo to Kandahar* (Abingdon: Routledge, 2009).

Michael J. Williams, *The Good War: NATO and the Liberal Conscience in Afghanistan* (Basingstoke: Palgrave, 2011).

Susan Woodward, *Balkan Tragedy: Chaos and Dissolution after the Cold War* (Washington, D.C.: The Brookings Institute, 1995).

Warren Zimmerman, *Origins of a Catastrophe* (New York: Random House, 1999).

Index